Waste Management

Kommission der Europäischen Gemeinschaften
Commission of the European Communities
Commission des Communautés Européennes

WASTE MANAGEMENT

Europäische Konferenz für Abfallbehandlung
European Conference on Waste Management
Conférence Européenne sur la Gestion des Déchets

Wembley, England, June 17-19, 1980

edited by

JEREMY WOOLFE

SPRINGER-SCIENCE+BUSINESS MEDIA, B.V.

Library of Congress Cataloging in Publication Data

European conference on Waste Management (1980: Wembley, London, England)
    Waste Mangement.

        At head of title: Commission of the European Communities. Kommission
der Europäischen Gemeinschaften. Commission des Communautés Européennes.
        1. Refuse and refuse disposal–Congresses. 2. Factory and trade
waste–Congresses. I. Woolfe, J. (Jeremy) II. Commission of the European
Communities. III. Title.
TD785.E89   1980   363.7'28   81–8681
ISBN 978-94-010-9944-8   ISBN 978-94-010-9942-4 (eBook)   AACR2
DOI 10.1007/978-94-010-9942-4

Publication arrangements by
Commission of the European Communities
Directorate-General Information Market and Innovation
EUR 7277 EN
Copyright © 1981 Springer Science+Business Media Dordrecht
        Originally published by ECSC, EEC, EAEC, Brussels and Luxembourg in 1981

LEGAL NOTICE

Neither the Commission of the European Communities nor any person acting on behalf of the
Commission is responsible for the use which might be made of the following information.

CONTENTS

NOTE TO READERS

WASTE MANAGEMENT AND RECYCLING, WHICH AS ASPECTS
OF THE "ENVIRONMENT" STARTED TO BECOME SERIOUS
POLITICAL ISSUES IN THE EARLY 1970s, COULD BY THE
80s HAVE BEEN GENTLY FADING AWAY.

HOWEVER IN EUROPE THE MOVEMENT IS GAINING GROUND,
NOTABLY IN THE EEC BECAUSE OF ITS HEAVY DEPENDENCE
ON IMPORTED COMMODITIES.

CONSIDERABLE DRIVE IS BEING GIVEN BY THE COMMISSION
OF THE EUROPEAN COMMUNITIES, WHICH IN THE ENVIRON-
MENT FIELD IN GENERAL HAS BEEN BEHIND 50 RELEVANT
DIRECTIVES, DECISIONS OR RECOMMENDATIONS INSTIGATING
A NUMBER OF BINDING RULES FOR INDIVIDUAL COUNTRIES.
THESE TOGETHER FORM THE BEGINNINGS OF A COHERENT
POLICY.

IT WAS IN THIS OVERALL CONTEXT THAT THE EUROPEAN
CONFERENCE ON WASTE MANAGEMENT WAS SPONSORED BY THE
EUROPEAN COMMUNITIES AND ORGANISED BY THE BRITISH
GOVERNMENT. ITS SUCCESS IS INDICATED BY THE ATTENDANCE
OF NEARLY 400 PEOPLE FROM ACROSS EUROPE.

FOR THE PURPOSE OF THIS PUBLICATION THE TEXTS OF THE
PAPERS DELIVERED AT THE CONFERENCE HAVE BEEN EDITED
AND MANY HAVE BEEN REDUCED IN LENGTH. EVERY CARE
HAS BEEN TAKEN IN PRESERVING AS MUCH OF THE ORIGINAL
THOUGHT OF THE SPEAKERS AS POSSIBLE BUT NO
RESPONSIBILITY IS TAKEN FOR ANY INADVERTENT
INACCURACY.

JW - 1980

INTRODUCTION

The EEC's heavy dependence on imported raw materials, combined with
insufficient priority given to recuperative technologies, provided twin
incentives to convene the European Conference on Waste Management.

The conference was also held in answer to a call for a serious international forum to enable high calibre spokesmen to express their views on a
diversity of vital interests.

The ambitious objective was that such a melting pot would smooth the road
to cohesive policies and purposeful action, especially in the field of
waste recycling.

Among the speakers there were several specialists ready to outline the
state of the art of a variety of technical subjects. This range included
aspects of packaging, hazardous waste, the utilisation of waste in agriculture, energy from waste, and waste paper.

Paper was to play a particularly serious role, because pulp and paper is
the largest external commodity need by the EEC countries after oil.

Speakers were to be drawn from all parts of the EEC, in the expectation
that country-to-country comparisons would stimulate progress in specific
areas. This hope was frequently realised.

Well represented was the Environment and Consumer Protection Department
of the Commission itself. It is this department that has been quietly
setting the beginnings of a coherent policy for waste management for
throughout the 250 million population Community for the past seven years.

Sokesmen were also to be drawn from sections likely to have quite
opposing views. Represented at the conference were to be local authorities and national government departments, both conscious of their responsibility to weigh the expenditure of their tax payers' money against the
long-term ecological interest of humanity.

There were also to be outspoken members of ecological groups, with stoutly

held philanthropic views, and on the other hand, members of several major
industries, with financial responsibility for the interests of their
shareholders and their employees to bear in mind.

All were ready to contribute their point of view. All had perfectly
valid statements to make. All were qualified to express themselves
eloquently.

In addition, and very much to the point, all were ready to listen to each
other ... and to attempt to rationalise and seek harmony from what could
have otherwise been a boiling sea of useless confusion.

In the event, as the concluding conference debate reported at the end of
this publication reveals, the final results were some refreshingly clear
indications of feasible policy lines to be advocated in the field of
waste management.

WASTE MANAGEMENT IN THE EUROPEAN COMMUNITY

Michel Carpentier

Head of the Environment and Consumer Protection Service
Commission of the European Communities

SYNOPSIS

It is in the environmental field that over the last seven years the
Community has made some of its biggest progress. With around 50
directives, decisions or recommendations, we have seen the beginnings
of a coherent European policy on the environment.

Such a policy is needed because many ecological problems, such as river
pollution, have to be tackled on an intra-Community basis; individual
solutions, if not harmonised throughout the Community, can lead to market
distortions and a common environmental policy enhances the position of
member states in the international arena.

The straightforward economic point is that the Community has to import
56 per cent of its energy (90 per cent of oil), 50 per cent of its paper
and wood pulp, and 80 to 90 per cent of metals. The total value of
"wasted" waste every year in the Community is 10 billion EAU, which, if
recycled, could give savings of 5 to 7 billion EAU.

Prevention of waste and the use of "clean technology" in product design.
European Symposium on Clean Technology at The Hague in November. Main
obstacle to recycling not financial but consumer resistance. Possibility
of setting standards of quality for secondary materials.

Waste oil and packaging, particularly beverage containers. Proposal for
a directive to limit the use of throw-away containers, and to encourage
returnable containers and recyclable materials.

Recommendation on recovery and re-use of waste paper to encourage the use,
particularly in the public services, of recycled paper and carton.

Toxic waste : transport and concentration of toxic matters in waste.

Forthcoming white paper on incineration and waste for energy purposes.

RESUME

C'est dans le domaine de l'environnement que la Communauté a accompli certains de ses plus grands progrès au cours des sept dernières années. Une cinquantaine de directives, de décisions ou de recommandations ont marqué les débuts d'une politique européenne de l'environnement cohérente.

Une telle politique est nécessaire en raison des nombreux problèmes écologiques, comme la pollution fluviale, qui doivent être résolus au niveau intra-communautaire ; en l'absence d'harmonisation au sein de la Communauté, les solutions individuelles risquent de conduire à des distorsions du marché, alors qu'une politique commune de l'environnement renforce la position des Etats membres sur le plan international.

La réalité économique est que la Communauté doit importer 56% de l'énergie (90% du pétrole), 50% du papier et de la pâte à bois ainsi que 80 à 90% des métaux dont elle a besoin. Les déchets "gaspillés" chaque année dans la Communauté représentent une valeur totale de 10 milliards d'UCE ; recyclés, ils permettraient de réaliser une économie de 5 à 7 milliards d'UCE.

Prévention de la formation de déchets et utilisation de "techniques propres" dans la conception des produits. Symposium européen sur les technologies propres à La Haye en novembre. Le principal obstacle au recyclage n'est pas d'ordre financier mais réside dans la résistance des consommateurs. Définition possible de normes de qualité pour les matières secondaires.

Gaspillage de pétrole et des conditionnements, notamment des récipients à boissons. Proposition de directive ayant pour but de limiter l'emploi de récipients à jeter et de promouvoir les récipients consignés et les matières recyclables.

Recommandation portant sur la récupération et le réemploi du vieux papier en vue d'encourager l'emploi, notamment dans les services publics, de papier et de carton recyclés.

Déchets toxiques : transport et concentration des matières toxiques
contenues dans les déchets.

Un livre blanc doit être publié prochainement sur l'incinération et
l'utilisation des déchets à des fins énergétiques.

Zusammenfassung

Einige ihrer grössten Fortschritte hat die Gemeinschaft in den ver-
gangenen sieben Jahren im Umweltsektor erzielt. Etwa 50 Richtlinien,
Entscheidungen und Empfehlungen stellten den Beginn einer zusammen-
hängenden europäischen Umweltpolitik dar.

Eine solche Politik ist notwendig, weil viele ökologische Probleme,
z.B. die Verschmutzung der Flüsse, auf einer innergemeinschaftlichen
Grundlage bewältigt werden müssen. Einzellösungen, die nicht inner-
halb der Gemeinschaft harmonisiert worden sind, können zu Verzerrungen
auf dem Markt führen; eine gemeinsame Umweltpolitik hingegen verstärkt
die Stellung der Mitgliedstaaten auf internationaler Ebene.

In wirtschaftlicher Hinsicht ist zu berücksichtigen, dass die Gemein-
schaft 56% ihrer Energie (90% des Erdöls), 50% ihres Papiers und
Zellstoffs sowie 80-90% der Metalle einführen muss. Der Gesamtwert
der"verschwendeten"Abfälle beläuft sich in der Gemeinschaft jährlich
auf 10 Milliarden ERE, die bei einer Wiederverwertung zu Einsparungen
von 5 bis 7 Milliarden ERE führen könnten.

Verhinderung von Verschwendung und Anwendung einer "sauberen Technologie"
in der Produktgestaltung. Europäisches Symposium über saubere Technologie
in Den Haag im November. Grösstes Hindernis gegenüber der Wiederaufbe-
reitung nicht finanzieller Natur, sondern Widerstand seitens der Ver-
braucher. Mögliche Aufstellung von Qualitätsnormen für Hilfsstoffe.

Altöl und Verpackung, insbesondere Getränkeverpackungen. Vorschlag
für eine Richtlinie zur Begrenzung der Verwendung von Wegwerfpackungen
und zur Förderung von Mehrwegverpackungen sowie von wiederaufbereit-
barem Material.

Empfehlung über die Wiederaufbereitung und Wiederverwendung von Altpapier, um insbesondere im öffentlichen Dienstleistungssektor den Ge-
brauch von wiederaufbereitetem Papier und Karten zu fördern.

Giftige Abfälle: Transport und Konzentration von giftigen Abfallstoffen.

Abfallbeseitigung - Weisspapier über: Verbrennung und Verwendung von
Abfällen zur Energiegewinnung.

When people think about the European Community it tends to be in terms of the Common Market and the Common Agricultural Policy.

In fact it is in the field of environment that the Community has, over the last seven years, made some of its most important advances.

It was in 1971 that the Commission of the European Communities first pointed out the need to take the environment into consideration when defining and organizing the economic development of the Community, and outlined Community action in this field.

This initiative, which was followed up by several Member States and acceding states, was given a favourable reception at the Paris Summit in 1972 and culminated the following year in the adoption of a Community policy and programme on the environment. The policy and the programme were reaffirmed and updated in 1977.

Since the adoption of the first programme, some fifty directives, decisions or recommendations relating to the environment have come into force and rules binding for the Member States have likewise been established. In these last seven years we have in fact seen the beginning of a coherent Community policy on the environment.

Quite apart from the fact that there is alot to be said for working together as a Community to solve problems that are common to us all, there are four very good reasons why the Community should have its own environmental policy.

One is that pollution knows no political boundaries and what happens in one country can have far-reaching effects in another. Europe, with its closely-knit pattern of states, is particularly exposed to these problems.

Why should the Dutch impose stringent quality objectives on their own industries with a view to keeping the Rhine river clean if Switzerland, France and Germany use it as a convenient dumping ground for their industrial wastes ? And why should Dutch and French factories have their dumping at sea severely restricted when the fish are, in any case, contaminated by waste substances from the British coast ?

Secondly, as a Community, we have set ourselves economic goals which have important implications for the environment. Material prosperity based on economic growth does not necessarily lead to a better quality of life. In other terms production and consumption should not be allowed to lead in turn to higher pollution, more waste and a rapidly deteriorating environment.

A third reason for uniform policy is that joint action in the environmental field is of paramount importance if distortion of competition and trade is to be avoided.

Finally, a common environmental policy enhances the position of the Member States on the international scene.

Take an obvious ecological problem : resource management, i.e. energy conservation, materials recovery and the control of waste. The disposal, recycling or re-use of waste depend on economic factors.

If the solutions adopted in one region or even one nation give rise to differences in the production and distribution conditions of certain goods, these differences may have repercussions on the functioning of the Common Market, and on international trade.

The waste problem

Wastes, of all kinds, are just one of the contradictory manifestations of economic growth. They may also be considered as "indicators" of the wastage of resources.

One aspect of the waste problem relates more specifically to environmental protection, that is the threat of pollution and nuisances caused by wastes, be they of industrial, domestic or agricultural origin. Costly treatment processes are often required to counter these risks.

The second aspect relates to the failure to utilize the materials contained in wastes. Changing consumer habits and developments in product design have created a situation where the amount of valuable materials contained in waste, such as paper and plastics, is steadily increasing.

Most of these materials have an obvious strategic and economic importance - globally in the long term - but, more immediately, for the Community. Remember : 33 000 Kcal of energy are needed to produce 1 kg of polystyrene; 14 000 Kcal are needed to produce 1 kg of paper.

The Community's dependence on non-member countries for its raw material and energy requirements are well known. The Community imports 56 per cent of its energy requirements - 90 per cent in the case of oil - 50 per cent in its needs in paper and woodpulp and 80-90 per cent of its requirements of iron and non-ferrous metals, particularly tin and zinc.

From the economic and strategic points of view, this situation has very serious implications : uncertainty over prices and supplies, imbalance of trade, unemployment, international problems.

Viewed worldwide, economic growth cannot continue on the basis of technologies and consumer habits characterised by a high consumption of energy and raw materials.

As to the present scale of the problem, the estimated quantity of waste produced in 1977 in the Community was 1 800 million tonnes, equivalent to nearly 5 million tonnes per day.

Moreover, the stream of waste is reckoned to be increasing by 2 to 3 per cent annually. The total value of unrecovered waste generated each year in the E.E.C. is probably greater than 10 000 million EUA. If these waste materials could be recycled, savings of between 5 and 7 million EUA would be achieved.

Thus the true nature of the waste problem becomes apparent : it is not merely a question of "managing" wastes and rendering them harmless. The whole question of the generation of waste must be reappraised and all possible ways found of re-using it.

As with the energy problem, the problem of waste - and, more so that of raw materials - is not automatically solved by market forces. The need for an effective policy by the public authorities in this field is therefore clear.

The objectives of the Community policy on waste and the reasoning behind it are expanded in Title III, Chapter 3, Section 2 point 178 of the 1977 Programme : "The protection of the environment against pollution, sound economic managemement of resources, the effort to reduce the Community's dependence on imported raw materials, the rational long-term management of natural resources which are either non-renewable or can be renewed at only a limited rate - all these considerations together argue in favour of an immediate and hard-hitting campaign against waste".

Three areas of action were designated in the 1977 programme :

  (i)   prevention of waste generation,
  (ii) safe disposal of non-recoverable wastes,
  (iii) recycling and re-use of wastes.

It was these three themes that the Commission had taken as the basis for a framework Directive 75/442/EEC on waste approved by the Council in 1975. The definition of objectives and the formulation of a general policy in the field of waste management is particularly complex.

For example, the degree of importance which should be attached to each of the three aspects mentioned - prevention, disposal recycling - depends on a number of technical and economic considerations. This difficulty led the Commission simultaneously to formulate the definition of a general policy and to take certain measures in respect of significant cases.

Prevention : clean technologies

Clearly manufacturers must be encouraged to make the most rational use of raw materials and energy and to reduce the amount of polluting effluent discharged into the environment and the quantity of waste produced during the manufacture and utilization of products.

Clean technologies contribute directly to this objective. The concept of "clean technology" covers three separate yet complementary objectives :

  -     to pollute less, i.e. discharge less effluent into the natural environment (water, air, soil) ;
  -     to produce less waste; this is why there is often talk of "low-waste" or "no-waste technologies" ;
  -     to be more economical in the use of natural resources (water, energy, raw materials).

The role of clean technologies is both quantitative, namely to reduce the waste stream and the amount of pollution and to save resources, and qualitative, namely, to change the composition of wastes and the type of pollution caused. Clean technologies may be brought in at either the design stage or at the production stage : attempts can be made to reduce the amount of waste or effluent produced (or even eliminate them alto-gether).

The Concept of clean technology is much broader than that of the design and operation of a production process. For this reason, it may be said to have a bearing on :

1. Product design (to ensure, for example, the maximum product durability, and to facilitate recovery and recycling after use).

2. Plant design and location (to make it easier for one plant to use the waste or energy produced by another).

3. Organization (measures affecting the organization of the plant and, in particular, the sequence of manufacturing operations can improve the effectiveness of the technologies used).

Apart from the regulatory action required, the public authorities may - in those sectors where pollution is must acute - promote the design and use of clean technologies by a range of measures, in particular :

- research and development projects,
- exchange of scientific and technical information,
- incentives for industrial cooperation (waste exchanges, technology transfer between industries (patents, licences, know-how) ,
- using Community or national aid to encourage industry to swtich over to clean technologies.

Among action planned by the Commission is a comparative survey of programmes undertaken by the various Member States in respect of the clean technologies. In addition, the Commission is to organize a European Symposium on Clean Technologies in The Hague in November 1980 in collaboration with the Netherlands Ministry of Health and the Environment.

Already an initial exchange of views with the European Council of Chemical Manufacturers' Federations has produced a greater degree of understanding of the definitions of clean technologies specific to the chemical industry.

The environmental research and development programmes include a topic devoted to clean industrial technologies. A call for the submission of proposals on this topic was published in the Official Journal of 2 November 1979.

At present, there are no regulations specifically aimed at promoting clean technology. A number of existing provisions might, however, encourage action along these lines. Principal among these is the Directive on waste of 15 July 1975.

Furthermore, a sectoral policy is beginning to evolve; it could make use of Directives such as the one relating to waste from the titanium dioxide industry (Directive of 20 February 1978) and give them greater force, for example as regards pollution emission limits.

This task is in line with the programme approved by the Council on 17 May 1977; this states that technical or other processes must be found for each polluting branch of industry which can reduce, eliminate or prevent the emission of polluting substances or the creation of nuisances.

These industrial sectors are sensitive not only because of the high degree of pollution they create and the corresponding effort they must take to reduce this pollution, but also by virtue of the effect of international trade on their development. One such industry currently under investigation is the paper and pulp industry. It is also proposed to investigate other sectors such as the chemical and food industries.

The granting of financial aid by the Member States under conditions which are to be harmonised at Community level could be contemplated under the auspices of the "European Conventions on the Quality of Life" which were discussed at the Council meeting of 9 April 1979.

It is important that any disparities in the aid granted by Member States to promote the clean technologies do not affect the conditions of competition nor the proper functioning of the common market.

The main objectives of action to further the recovery of materials has to be aimed at the re-use of the products and/or materials contained in wastes. Industry already recovers and re-uses a number of materials and this practice must be extended.

As to municipal refuse, there are two alternatives : the materials or products can either 1) be separated at source which require above all an organizational effort and the cooperation of households or 2) be separated at a waste treatment centre.

The objectives and the attractiveness of recovery from the point of view of resources and environmental protection are quire clear; in order to coordinate and guide the work being done, the cost of the operation and the obstacles to be overcome must be investigated.

The following conclusions may be drawn from studies carried out for the Commission :

> (1)   the recycling of waste - where technically feasible - is generally a better financial proposition than to use waste to produce fuel or compost ;
>
> (ii) as far as possible, the recovery of waste materials should be carried out at source ;
>
> (iii) much greater use must be made of residues which cannot be recycled (e.g. the use of furnace clinker in road building). At present there is little financial incentive for such practice ;
>
> (iv) the cost of recovery is highly dependent on the local situation (transport costs, waste stream etc.) ;
>
> (v)   certain types of treatment to upgrade recovered materials can make them competitive with the corresponding raw materials.

The general finding is that there are several financially attractive recovery techniques. There are, however, a number of obstacles to the widespread introduction of recycling. Industry and consumers tend to prefer non-recycled materials and products.

The result is that secondary raw materials are in demand only in periods of high output, the demand falling rapidly as soon as the raw material supply difficulties disappear. Consequently the market lacks stability, giving no incentive to invest.

Recycling should therefore be encouraged by the parallel development of supply and demand for recycled materials and products and by stabilizing the pattern of demand. Any unilateral increase in supply would have the effect of reducing prices and thus discourage those firms interested in recycling.

A range of measures to combat the problems may be contemplated, namely :

- incentives to promote the use of specific recycled materials and to discourage the use of the corresponding raw materials ;
- guidelines setting out the conditions under which recovery and recycling should take place ;
- fixing quality standards for "secondary" materials to provide guarantees for potential users and to eliminate any unfair discrimination ;
- specifying the recycled materials content in the products purchased or in the service contracts concluded by government departments ;
- encouraging research into recovery and recycling.

Among a number of steps already taken by the Community in key areas is a Council Directive of 16 June 1975 (OJ L 195/75) that provides for a regulatory framework for the disposal and regeneration of waste oils. Since such oils are a prime cause of water pollution, a strict system of controls has been set up covering the collection, dumping, disposal and/ or regeneration of such oils.

The total consumption of lubricants in the Community is approximately 4 million tonnes per year and the amount of waste oil generated annually is 2 million tonnes, of which only just over one half is collected and the rest disposed of without control.

A more detailed study must still be made at Community level of the comparative merits of regeneration and combustion. Combustion without pre-treatment increased since 1973, but treatment is needed to remove heavy metals which otherwise cause atmospheric pollution.

Another case investigated is packaging, which represents about 30 per cent by weight of urban refuse. The production of packaging requires a considerable input of energy and raw materials. Packaging also causes litter and contributes to air and water pollution.

Because all social costs involved (some 680 million EUA) are not borne by those responsible, i.e., the manufacturers, distributors and consumers, market forces are not available to prevent the achievement of an optimum balance between the various types of packaging.

The Commission has been studying the problems caused by beverage containers - a key outlet for the packaging industry - which accounts for some 10 per cent by weight of urban refuse.

A proposal for a Council Directive on beverage containers aims to limit the use of throw-away containers and to introduce incentives to return the containers to the point of sale and to recycle the materials contained in the various types of packaging.

The recycling of waste paper is another key feature of waste management in the Community. Waste paper represents about 15-20 per cent by weight of urban refuse (15-17 million tonnes per year) and between 30-50 per cent by volume. With a shortfall to the value of 8 000 million EUA, wood and its derivatives are in second place behind oil products.

At present, between 9 and 10 million tonnes of waste paper are used in the manufacture of paper and board; this represents a third of the Community's consumption (30 million tonnes). By using waste paper and not raw pulp, energy consumption is cut by a factor of six, and pollution is reduced. There are, however, a number of constraints on the use of waste paper which are both economic and technical.

A draft recommendation on the recovery and re-utilization of waste paper and board was recently submitted by the Commission to the Council. In essence, its adoption would encourage :

- the use - especially by the public services - of recycled and recyclable paper and board ;
- the use of recycled paper with a high mixed waste paper content ;
- the re-utilization of waste paper by other means.

Disposal is to define the conditions under which waste disposal should be carried out. The most important example is that of toxic and dangerous wastes. Directive 78/319/EEC (OJ L 84/78) lays down rules for the collection, storage, transport and treatment of these wastes by a system of registers and operating licences, together with monitoring of national programmes.

The need to take adequate precautions in respect of collection, disposal and re-utilization of toxic waste is obvious. The Directive lists 27 substances in respect of which precautions are vital. Noteworthy among these are mercury, cadmium, certain solvents and pharmaceutical compounds, ethers, tar and asbestos.

Work is currently being carried out in collaboration with the Member States in the areas of transport of wastes and the concentration of toxic and dangerous substances in such wastes. There is agreement that the potential hazard presented by these wastes must be assessed in terms of their concentration of toxic and dangerous substances. An inventory must therefore be drawn up of the waste streams and of existing treatment plants from both the quantitative angle (volume of the wastes and plant capacities) and the qualitative angle (type of waste produced and the treatment capability of any given plant).

An especially dangerous category of wastes is that of the polychlorinated biphenyls (PCB) and the polychlorinated terphenyls (PCT). Directive 76/403/EEC regulates the disposal of such substances, its main aim being to prohibit the uncontrolled discharge, dumping and tipping of these wastes and to make their safe disposal compulsory. The establishments responsible for the disposal of PCBs are subject to a system of prior authorization. The Directive also stipulates that the Member States shall promote the recovery of PCBs.

One particularly interesting method of waste disposal is incineration. The Commission is shortly to publish a paper on incineration and more generally on energy from waste in which it will describe the results of its work in this field.

Conclusions

The 1977 action programme on the environment stresses the need for an active campaign against waste. The Council of Ministers of the Environment has asked the Commission to continue its work with a view to assessing the scale of the problem for the Community, to compare the action already taken by the Member States, and to put forward proposals relating, in particular, to the ways in which the public authorities can intervene in specific sectors.

The Commission is to submit a discussion paper to the next meeting of
the Environment Council on implementation of the action programme.
This document highlights the problems encountered.

A major problem in establishing a waste management policy stems from the
diversity and multiplicity of the experiments carried out at widely
different levels of responsibility, as a result of which numerous
contacts are required. Such contacts, however, are a prerequisite for
any exchange of information.

They require a minimum number of staff at a time when the staff shortage
is particularly acute. It has not been possible, for example, to
convene twice yearly the Waste Management Committee — set up in 1976
by the Commission to assist it in this field — and at the same time
draw up proposals to be put to this Committee. Nevertheless, a
Community policy on waste management is needed.

Whereas certain aspects of the problem are of a specifically local
nature (availability of tipping sites for example), it must be said
that the way in which the basic problem manifests itself is identical
in all the industrialised countries.

Major savings can be achieved at Community level by coordinating
research or carrying out pilot projects. Given the increasing inter-
penetration of the national economies, Community-wide harmonization of
the regulations governing the various aspects of the waste problem is
needed in order to prevent discrimination and market distortion.
Furthermore, a policy of incentives should be framed at Community level.
The creation of a new Community financial instrument for environmental
protection is currently under consideration.

It is not appropriate at this stage to draw up a list of strict
priorities for this policy. In the medium term priority should be given
to the development of clean technologies. In the first place, effort
must be directed at creating the conditions for technological change
compatible with economic and social requirements.

An urgent effort should be made in the short term to expand the policy
on waste recovery and to limit the most damaging effects of present
consumer habits. Finally, detailed and strict conditions must be laid
down governing the disposal of non-recoverable residues.

A campaign to promote greater public awareness is therefore needed :
the fact must be accepted that a certain transfer of resources must take
place if there is to be a flexible and effective waste management policy.
I am certain that our conference will help to promote this awareness.

This conference has brought together not only representatives of the
public authorities concerned with waste management but also
representatives of the various economic sectors involved. It is hoped
that it will broaden our understanding of waste management and will in
addition enlighten the Commission as to its future action.

THE CHALLENGE OF RECYCLING

Rt. Hon. Tom King
Minister for local Government and Environmental Services
Great Britain

SYNOPSIS

Finite limit to most of world's resources. At present rates of consumption reserves of iron will last 93 years, aluminium 31 years, copper 21 years, zinc 18 years, tin 15 years. Hence reclaiming materials from waste makes sense, but often the process is uneconomic. But price structures could change, having dramatic effect on economies of recycling.

Need to keep on with research and development to meet challenge. Welcomes work by European Communities. Recycling commercial waste tends to be successful. Municipal waste - large sums being spent on engineering know-how on waste separation. Three plants in U.K. Clean technology - need to rely on persuasion rather than compulsion.

Returnable beverage containers - refers to draft directive and to U.K. working group. Welcomes draft recommendation on waste paper.

RESUME

La plupart des ressources existant dans le monde s'épuiseront. Au rythme de consommation actuel, les réserves de fer dureront 3 ans, celles d'aluminium 31 ans, de cuivre 21 ans, de zinc 18 ans, d'étain 15 ans. La récupération de matières à partir des déchets est donc une bonne idée, mais il arrive souvent que les procédés ne soient pas rentables. Les structures de prix pourraient néanmoins changer et se répercuter de manière spectaculaire sur l'économie de recyclage.

Il est nécessaire de poursuivre les activités de recherche et de développement pour relever le défi. L'auteur se félicite du travail accompli par les Commonautés européennes. Le recyclage des déchets commerciaux s'annonce comme une réussite. En ce qui concerne les déchets urbains, des crédits importants sont affectés à la mise au point de méthodes de séparation des déchets. Il existe trois usines au Royaume-Uni. L'introduction de techniques propres est affaire de persuasion plutôt que de contrainte.

Pour ce qui est des bouteilles consignées, l'auteur mentionne le projet de directive et le groupe de travail du Royaume-Uni. Il se félicite du projet de recommandation sur les vieux papiers.

Zusammenfassung

Die meisten Ressourcen der Welt sind begrenzt. Bei dem derzeitigen Verbrauchstempo reichen die Eisenreserven für 93 Jahre, die Aluminium-reserven 31 Jahre, die Kupferreserven 21, die Zinkreserven 18 und die Zinnreserven 15 Jahre. Daher ist die Forderung nach Material aus Ab-fällen berechtigt. Das Verfahren ist jedoch oft unwirtschaftlich. Preis-strukturen können sich jedoch verändern, was erhebliche Auswirkungen auf die Einsparungen durch Recycling haben würde.

Notwendigkeit der Fortsetzung von Forschungen und Entwicklung, um die Herausforderung anzunehmen. Die Arbeiten der Europäischen Gemeinschaft sind willkommen. Die Wiederverwendung von Industrieabfällen hat immer mehr Erfolg. Städtische Abfälle - grosse Summen werden für die tech-nischen Kenntnisse über die Trennung der Abfälle aufgewandt. Drei An-lagen im Vereinigten Königreich. Saubere Technologie - Notwendigkeit, sich mehr auf Überzeugung als auf Zwang zu stützen.

Einweggefässe für Getränke - Bezugnahme auf Richtlinienentwurf und Arbeitsgruppe des Vereinigten Königreichs. Empfehlungsentwurf auf Abfallpapier willkommen.

When thinking of waste management, one automatically turns to waste reclamation and recycling. There are two reasons – it is increasingly difficult to find suitable holes in the ground to dispose of ever growing mountains of waste – but more importantly, we must preserve valuable materials, some of whose resources are rapidly running out or, even if more plentiful, have to be imported at the cost of valuable foreign exchange.

Over the last decade, the whole of the developed world has woken up to the fact that there is a finite limit to most of the world's resources and that we have been using them at a rate that is unsustainable. Just a few examples : experts estimate that, at projected rates of consumption, reserves of iron will run out in 93 years, aluminium in 31 years, copper in 21 years, zinc in 18 years and tin in 15 years. Of course, this is one view. Others maintain that as the amount of resources dwindles, prices will rise sufficiently to ensure exploration takes place to find new sources. What nobody disputes is that these resources are ultimately finite. The only argument is about the time-scale.

In these circumstances there can be no doubt about one thing – to reclaim materials from waste and use them again makes undeniable sense. But to admit this fact is only the start. Time and time again, we run into the same situation. We do our sums – we work out what it costs to collect the waste materials, to assemble them, to transport them to the location of the industry that is to use them and to prepare them so that they can be recycled. And then we find that the whole process is uneconomic.

Now at times of world recession, when all our economies are at the best stagnant, it does not seem to make sense to undertake reclamation and re-cycling operations just to be "good guys" if we know it is losing us a lot of money. So the challenge of recycling in the 1980s is to devise means of making the whole process economic and therefore really worthwhile.

What are the possibilities ? One of the basic things we should ask our-selves is whether price relativities for finite raw materials will remain as they are. Consider what has happened to oil. It has increased from $ 2 a barrel in 1973 to about $ 32 a barrel now. This could well happen to other commodities, when it becomes evident that they are in palpably short

supply. There is little doubt that - if that happened - the developed
world would soon learn to adjust. There would be a rapid chase for
alternatives and economies in use. But it would have a dramatic effect
on the economics of recycling. So we must be ever alert for the possibilities - what may not be economic today may very well be extremely
worthwhile tomorrow.

We must keep on with our research and development in order to be in a
position to meet every challenge. It is through the efforts of our
scientists and engineers that we shall find new ways of separating valuable materials from the various types of waste - industrial, commercial
and domestic. And where we already know how to do it, how to develop
and refine the process until it becomes possible to carry it out economically on a large scale. Here, I welcome the work done by the European
Communities in their various research and development programmes - programmes which work for the benefit of us all and enable us to profit
greatly from each other's experience.

Next we must look at markets for reclaimed materials. In my experience
there is a great inertia on the purchasing side of industry and reluctance
to lower specifications when it would be perfectly reasonable to do so.
This is where we must undertake a programme of education to ensure that
where opportunities are economic, they are appreciated by those concerned.

DOMESTIC AND COMMERCIAL WASTE
There is little doubt that reclamation and recycling in the field of industrial waste tends, with the help of the research and development I have
mentioned, to be more successful than in the field of domestic and commercial waste. Since waste disposal costs for industry are high there is
more incentive, the materials to be recovered are often more valuable and
they tend to arise in much greater concentrations.

So reclamation and recycling from domestic and commercial waste presents
a greater challenge. Separated collections of different materials from
individual homes and commercial premises seem these days to be out of the
question. We are having some success with schemes in which the public
bring materials to central collection points, but we are pinning great
hopes on the development of mechanical sorting plants.

The Government has invested large sums of money and much engineering know-how in the development of these plants. Now, with the cooperation of local authorities involved, there are three plants operating - a large scale test plant at Chichester, and full scale plants at Newcastle and Doncaster. These plants are designed to reclaim automatically various materials from the waste and with the residue produce a pelletised waste-derived fuel.

The next 18 months will be a critical period for completing the projects and assessing their technical and economic viability. Naturally there are problems but bearing in mind the experimental nature of the projects, it is not felt that these are excessive. I am confident that we are getting them ironed out. We do now believe that we are on course for a viable waste disposal option for large conurbations which will also make a useful contribution to our energy resources.

There is a heavy responsibility on industry to design products so that they can be made with the minimum amount of scarce resources, so that they last longer and so that the various materials used in their manufacture can be easily separated at the end of their life. In short, what is known these days as "clean technology".

I know a lot of thought has been given to these kinds of ideas already, but it is difficult to persuade designers to assume their responsibility. In a free enterprise society - one which we would not want to change - it is impossible to achieve this objective by legislation (indeed legislation would be almost impossible to draft. Therefore, we must rely all the more on persuasion and customer pressure.

BEVERAGE CONTAINERS AND WASTE PAPER
I should just like to say a final word about the topical subjects of beverage containers and waste paper. Beverage containers are a subject of considerable importance. Thanks to our system of doorstep milk delivery, this country can claim the highest usage rate of returnable containers in the world. But we should not accept this situation without question. The European Community has been studying this whole area very actively and has produced a draft directive, which is now under consideration. At home, a working group has also been studying the whole question in depth and I

await its report with anticipation.

On waste paper, the Government warmly supports the recent draft recommendation sent to the Council by the Commission and I shall be considering how best we can further increase our efforts in this area.

I return finally to recycling. I repeat : the challenge is to make it
economic by every possible method. The dwindling of world resources gives
us no choice in the matter : we must keep this objective firmly and
permanently in view.

GUIDLINES AND PROSPECTS FOR COMMUNITY RESEARCH INTO

THE RECYCLING OF URBAN AND INDUSTRIAL WASTE

Dr Matteo Donato   Dr Gian L. Ferrero

Directorate-General for Research, Science and Education

Commission of the European Communities

SYNOPSIS

Following the Decision of the Council of the European Communities on 12 November 1979, the Commission is implementing a research and development programme on urban and industrial waste.

The programme is divided into four research fields while in turn are sub-divided into topics.

In each field, research financed by the public authorities in the Member States will be coordinated and supplemented by research projects on specific topics, partly financed by the Community in the form of indirect actions (contracts generally allocate expenditure on a fifty-fifty basis).

This document gives a brief description of the research topics covered by Community indirect actions and some of the reasons behind the choice of topics for the programme and of the Community's general policy. It also mentions the research topics which will have to be coordinated.

Some important statistics are given to justify the programme.

Finally, the document describes the EEC's present and future policies.

RESUME

Comme suite à la décision du Conseil des Communautés européennes du 12 novembre 1979, la Commission met en oeuvre un programme de recherche et de developpement dans le secteur des déchets urbains et industriels.

Ce programme est divisé en quatre secteurs de recherche qui se subdivisent eux-mêmes en thèmes de recherche.

Dans chacun de ces secteurs, il sera procédé à une coordination des travaux de rechercher financés par les autorités publiques des pays membres, qui sera complétée par des projets de recherche sur des thèmes spécifiques, financés partiellement par la Communauté sous la forme d'actions indirectes (contrats à frais partaæs, en général à 50 %).

Le présent document décrit brièvement les thèmes de recherche couverts par les actions indirectes communautaires ainsi que certaines des motivations qui ont conduit au choix du programme et des orientations générales de la Commission. Il indique en outre les thèmes de recherche couverts par les activités faisant l'objet de la coordination.

Certaines statistiques importantes mettent en évidence les motivations du programme.

Les orientations actuelles et futures dans la CEE complètent la presentation du document.

ZUSAMMENFASSUNG

Dem Beschluss des Rates der Europäischen Gemeinschaften vom 12 November 1979 zufolge führt die Kommission ein Forschungs- und Entwicklungsprogramm auf dem Gebiet der Haushalts- und Industrieabfälle durch.

Das Programm gliedert sich in vier Forschungsbereiche, die wiederum in Forschungsthemen unterteilt sind.

In jedem dieser Bereiche werden die von den Behörden der Mitgliedstaaten finanzierten Forschungsarbeiten koordiniert und durch Forschungsvorhaben über spezifische Themen ergänzt, die zum Teil als indirekte Aktionen von der Gemeinschaft finanziert werden (Kostenteilungsverträge, bei denen in der Regel 50 % der Kosten zu lasten der Gemeinschaft gehen).

Das vorliegende Dokument enthält eine kurze Beschreibung der Forschungs- themen in Rah men der indirekten Aktion der Gemeinschaft sowie einige der Gründe, die zur Wahl des Programms und der allgemeinen Ziele der Germeinschaft geführt haben. Ferner sind die Forschungsthemen erwähnt, auf die sich die Koordinierungstätigkeit bezieht.

Die Gründe für die Durchführung des Programms werden durch relevante statistische Daten verdeutlicht.

Die gegenwärtigen und die künftigen Zielsetzungen in der EWG ergänzen die Vorlage des Dokuments.

On 12 November 1979, the Council of the European Communities decided that, with effect from 1 November 1979, the Community would carry out a four-year research and development program on the recycling of urban and industrial waste (secondary raw materials). An appropriation of nine million European units of account has been earmarked for the programme.

This document briefly explains how the subjects were selected and summarizes the research topics covered and the results expected.

The remarkable scientific and technical advances of the last hundred years have made available a great abundance of food, raw materials and energy, than ever before. This has led to unprecedented population growth and rise in living standards which have a direct impact on world resources.

We now realize, although the light dawned only recently, that these resources are finite and that we should therefore ensure that they do not run out too soon. Accordingly, ever-increasing interest has been shown over the last few years in the problems of conserving energy and resources, finding substitutes and reclaiming and recycling materials which can be used again.

The reclamation and recycling of urban and industrial waste covered by this general program can make a considerable and economically attractive contribution.

In 1975, the volume of waste in the European Economic Community alone was estimated at some 500 million tonnes, or 4.2 million tonnes a day, including :

> 90 million tonnes of household waste, which works out at around
> 300 k/year per person ;

950 million tonnes of agricultural waste ;

115 million tonnes of industrial waste ;

200 million tonnes of sewage sludge ;

150 million tonnes of waste from the extractive industries.

These quantities are increasing at an average rate of around 3 per cent a year in the European Community, although there are differences from one Member State to another.

It would obviously be beneficial, therefore, to develop technologies which can maximize the amount of waste reused and reduce the amount of waste generated by today's industrial processes, particularly as these new reclamation methods could help to solve a growing number of ecological problems.

An essential corollary of reclamation however is the development of markets for the secondary materials it produces so that recycling operations can be economically viable.

Waste can thus be seen as a reserve of unused, or partially used, resources - in some cases an extremely important one - of which fuller use can and should be made in the future. The use of waste can, if carried out properly, be tantamount to conserve rare and essential resources both energy, and raw materials, with all the economic advantages this would bring.

In the last few years, public authorities, industry and scientists in the various Member States have shown a growing interest in reclaiming usable products from waste precisely in order to save raw materials and energy - and thereby reduce dependence on imports - and to safeguard the environment.

Raw materials supply problems have spurred the Community to carry out research and development programs on primary raw materials based on the work done by a special subcommittee of CREST (Scientific and Technical Research Committee) and other measures to ensure security of supply are planned.

The present research and development program on recycling urban, industrial and agricultural waste is the continuation of the Community programs on primary raw materials, uranium exploration and extraction and paper and board recycling, adopted by the Council in March and April 1978. The program was based on the work of the above-mentioned CREST subcommittee on raw materials research and development and its working party on household waste, and on the findings of a number of studies contracted out to specialized bodies in the Member States on the following topics :

    sorting at source of household waste ;

    mechanized sorting of household waste :

    thermal waste treatment processes ;

fermentation and hydrolysis of organic waste ;

recovery of rubber waste.

The findings of these studies are set out in seven reports containing a full and up-to-date analysis of current research and of the state of the art in these fields (see attached list).

The present program covers two of the four priority areas selected for the common policy in the field of science and technology in view of the general objectives of the Community, namely the long-term security of supply of resources (raw materials), and environmental protection and nature conservation.

The program also meets various general and special criteria for Community research, particularly that of improving the effectiveness of Community action by pooling national research resources and eliminating unnecessary duplication of effort, and at the same time meeting the common collective requirements of all the Member States.

R & D Program

The current four-year program comprises four research areas, which in turn are subdivided into research topics. In each of these areas, current research work financed by the public authorities in the Member States is coordinated and supplemented by research projects on specific topics which are partly subsidized by the Communities in the form of indirect action (contracts where costs are generally shareed 50-50). The topics are decided upon by the Commission on the basis of the opinion of the advisory committee of national officials ACPM which helps to manage the program.

For the sake of clarity, the program is subdivided into two parts :

Research carried out in the form of indirect action ;

National research to be coordinated at Community level.

There follows a brief summary of the reasons for the research topics in each part of the program and the aims behind them.

Research carried out in the form of indirect action takes the form of projects on specific topics partly subsidized by the Communities in the form of contracts where costs are generally shares 50-50.

Research into the recovery of materials and energy from urban waste is covered in the paper submitted by Mr Noto La Diega and is therefore mentioned only briefly here. This area is included in the Community program for a number of reasons :

the growing interest shown by national and local authorities and industry in techniques for reclaiming useful products from urban wastes in order to reduce the quantity of waste which has to be disposed of ;

to exploit alternative sources of raw materials ;

to develop disposal systems which are economically viable and ecologically safe.

This practice is thus a realistic way of reducing the volume of domestic refuse and the amounts to be composted or used as landfill. About 50 per cent of the urban waste collected in the Community is used as landfill, while an average of 21 per cent to 22 per cent is incinerated (20.5 million tonnes in 1977, about 60 per cent with energy recovery) and less than 10 per cent composted.

Technology for the sorting of bulk wastes can be classified into : air classification, comminution (liberation) and novel techniques. Study into air classification includes investigation of the factors which affect separation in order to facilitate the choice of air classification systems for use in special situations and also covers working out large scale tests concentrating particularly on the comparison of air classifiers under similar conditions ; the effects of changing the variables in the project and the composition of the waste ; the definition of criteria for the projects ; and gradually increasing the dimensions of air classification systems.

The objective of comminution (liberation) is to make comparative studies of comminution and liberation systems capable of treating 10 tonnes or more of waste per hour in similar conditions with different types of waste. This is to determine : the extent to which treated waste can be further separated into several products ; the disposal of particularly bulky waste ; and system reliability and maintenance problems.

Research work into novel techniques is to be based on :
special density methods ;

optical methods ;   and

ballistic, triboelectric and other methods.

To supplement the R & D program on paper and board recycling adopted by
the Council on 17 April 1978, attention is to be given to the design and
operation of air classifiers in order to obtain a product with a higher
paper content and improving techniques for separating plastic film from
paper.

Work in the field of plastics is to concentrate on the composition of
waste to obtain particular finished products, extraneous matter which
prevents the use of reclaimed polymers, and development of new additives
for moulding. It will also cover reclamation of plastic originating from
specific sources of consumer waste, such as airports, large hospital
complexes and so on.

As work is already being done in the energy section of the Community
research and development program, research topics in the present program
are to be concerned with only the preparation and treatment of waste-
derived fuel (WDF), whether crushed or separated by density.

They will also cover the use of this fuel, with particular regard to the
following points :

-       the extent to which burning WDF causes metal corrosion and the
        scaling of boiler pipes ;

-       discharges into the atmosphere of products arising from the
        combustion of WDF.

Consideration will be given to the possibility of financing work on new
systems for collecting and transporting urban waste, particularly the use
of underground pneumatic conveying systems.

The reason for including this area in the Community program is that the
thermal treatment of waste (incineration and pyrolysis) offers a number
of advantages over ordinary dispersal : it reduces the volume of waste,
can generate energy, is quick to operate, sterilized the residual waste,
and so on.

In 1977, the incineration of urban waste in plants with energy recovery
in the European Community saved energy equivalent to some 3.3 million
tonnes of coal (1.32 per cent of total coal consumption), or 2.2 million
tonnes of oil (0.42 per cent of total fluid hydrocarbon consumption).

Urban waste is incinerated on a large scale throughout the Community and a satisfactory technical level has been achieved. Nevertheless, different methods have to be used for the disposal of other types of industrial waste, e.g. pyrolysis and gasification. These technologies offer attractive possibilities for the disposal of particular kinds of waste and at the same time generate energy and produce secondary materials.

They would repay research aimed at optimizing the processes and making them more reliable, and analysing them on a suitable scale and under various conditions, to assess their potential and make accurate recommendations on their use. Studies on the research topics in this area of the program which are covered by indirect action should concentrate on the following points.

Research in the fermentation and hydrolysis area covers agricultural and forestry waste and organic industrial and urban waste.

Each year the farming sectors, the food-processing industry, forestry and households in the Community produce large quantities of organic waste (over 1 000 million tonnes). Although they may be used for landfill or be disposed of by burning, the use of fermentation and hydrolysis would enable useful organic products and energy to be recovered.

The aim of the studies on anaerobic digestion is to contribute towards the full development and acceptance of the process for the treatment of most agricultural and industrial waste.

The reasons for the studies on carbohydrate hydrolysis (cellulose) are that cellulose and starch of organic origin can be hydrolyzed chemically and enzymatically to give glucose which can be used as such or as a feed-stock for producing other useful chemicals.

The aim of studies in this field is thus to improve existing technology for this process and to make it more commercial by competition. Studies on the above-mentioned research topics which are covered by indirect action, should concentrate on the following points :

Anaerobic digestion

Development of simple, economic and reliable digesters which can be manufactured economically for the treatment of agricultural or industrial waste, with particular emphasis on speeding up digestion processes by, for example, mixing the waste with household waste.

- basic studies on anaerobic, microbiological and biochemical
  digestion including the production of various substances from
  methane, and thermophilic digestion.

Carbohydrate hydrolysis

pre-treatment processes to improve the sensitivity of cellulose to
hydrolysis.

- development of single-stage microbiological processes for
  various chemical products.
- enzymatic hydrolysis : selection of species and type noted for
  good enzymatic production, development of techniques using
  immobilized enzymes.
- recovery and use of lignin.

It is important that national research, that is research receiving no
financial contribution from the Commission, is coordinated at Community
level. The Commission is to coordinate research on household waste into
the assessment of waste sorting projects ; methods for sampling and
analysing household waste ; evaluation of health hazards, and the recovery
of non-ferrous metals.

It is also to coordinate the thermal treatment of waste and the recovery
of metals and glass from incineration slag and pyrolysis. Fermentation,
hydrolysis and composting are also research subjects to be coordinated.

In the field of the recovery of rubber waste, research and technological
experiments have already reached an advanced stage of development or
even the industrial demonstration stage. The possibility of direct action
in this field is to be reviewed in the future.

Research topics which to begin with are merely to be coordinated by the
Commission cover retreading, size reduction, reclaiming and recycling
rubber powder, and pyrolysis.

General research policies within the EEC are difficult to define, but
trends toxards improving recovery and recycling techniques have been noted
in the recovery and upgrading of materials and energy from household and
industrial waste by mechanical or thermal treatment methods, and the
upgrading of agricultural waste by converting it into products which can
be used by the agri-food and chemical industries.

Successful research can form a coherent whole and could bring invaluable practical benefits in overcoming raw materials supply difficulties in the medium term. It would also enable industry within the Community to export technology in recycling on a world scale.

EEC R & D PROGRAM FOR PAPER AND BOARD RECYCLING

Ing Edmund Fassotte

Directorate-General for Research, Science and Education

Commission of the European Communities

SYNOPSIS

1. *Aim*

Increase recycling of waste paper in the paper and board industry of the Community by upgrading the main types of waste paper.

2. *Motivation*

a) *To solve, or at least reduce, the raw materials supply problems of the paper and board industry of the Community.*

b) *To ensure better management of industrial and urban waste by efficient re-use of the fibrous matter.*

3. *Description of the Programme*

In order to re-utilize additional amounts of waste paper and board, it is necessary to upgrade all types of waste paper by means of new processes.

Four major research topics were selected to this effect; they cover most R & D needs in this area :

1st topic : Definition of reclaimed fibres, their upgrading by various processes and the effects of multiple recycling on paper-making fibres.

2nd topic : Elimination of the detrimental effect of contaminants in waste paper.

3rd topic : De-inking and the treatment of effluent from waste paper recycling plants.

4th topic : Use of urban fibres and health problems caused by the use of recycled fibres.

4. *Implementation of the programme*

The programme will be implemented in the form of indirect action by means of contracts financed partly by the Community Budget and concluded with private and public research organizations in the Member States. It was the subject of a Council Decision dated 17 April 1978 which fixes the maximum amount contributed by the Community at 2 900 000 EUA for a three-year period.

The Commission invited interested parties to submit projects and has concluded 19 contracts to date (10 January 1980). There are still 7 contracts to be concluded in the next few weeks.

The first results are expected at the end of 1980.

RESUME

1. But

Intensifier le recyclage des vieux papiers dans l'industrie communautaire des papiers et cartons par une valorisation des principales sortes de vieux papiers permettant une meilleure utilisation des différentes qualités de vieux papiers, entraînant une répartition plus judicieuse des matières premières papetières en fonction des produits fabriqués.

2. Motivation

a) Réduire sinon résoudre les problèmes d'approvisionnement en matières premières de l'industrie papetière communautaire. Le déficit en matières premières de l'industrie papetière communautaire est une des principales origines des difficultés rencontrées par cette industrie et une des causes de la faiblesse de sa capacité de production qui doit être compensée par des importations de plus en plus importantes de papiers et cartons.

b) Permettre une meilleure gestion des déchets industriels et urbains en assurant une réutilisation économique des matières fibreuses qu'ils contiennent en grandes proportions ce qui entraine une réduction proportionnelle des déchets à éliminer.

3. Description du programme

Sur le plan technique, il apparaît que pour réutiliser des quantités de vieux papiers et cartons supplémentaires, il faut revaloriser toutes les qualités de vieux papiers par des technologies et processus nouveaux et mieux appropriés.

Quatre thèmes principaux de recherche ont été retenus à cet effet et couvrent l'essentiel des besoins actuels de recherche et développement en ce domaine :

1er thème : Caractérisation des fibres recyclées, leur revalorisation par procédés divers et les effets du recyclage multiple de la fibre.

2ème thème : Elimination des effets nuisibles des contaminants dans les vieux papiers.

3ème thème : Désencrage et traitement des effluents des usines de recyclage des vieux papiers.

4ème thème : Utilisation de fibres urbaines et problèmes sanitaires résultant de l'usage de fibres recyclées.

4. Exécution du programme

Le programme est mis en oeuvre sous la forme d'une action indirecte par la voie de contrats financés partiellement sur le budget de la communauté, conclus avec des organismes de recherches privés ou publics situés dans les Etats membres. Il a fait l'objet d'une décision du Conseil en date du 17 avril 1978 par laquelle il bénéficie d'une contribution communautaire maximum de 2.900.000 UCE pour une durée de 3 ans.

Suite à appel d'offre, la Commission a pu conclure à ce jour (10 janvier 1980) 19 contrats. Il reste à conclure 7 contrats au cours des prochaines semaines.

Les premiers résultats sont attendus dès la fin de 1980.

ZUSAMMENFASSUNG

1. *Programmziel*

Intensivierung des Altpapier-Recycling in der Papier- und Pappeindustrie der Geienschaft durch verstärkte Rückgewinnung der wichtigeren Sorten von Altpapier, die eine bessere Wiederverwertung der verschiedenen Qualitäten des Altpapiers erlauben, mit dem Ziel einer vohgeplanten Verteilung der Papierrohstoffe je nach den Herstellungsprodukten.

2. _Motivation_

a) Verringerung und wenn möglich Lösung der Versorgungsprobleme der
   Gemeinschafts-Papierindustrie. Das Rohstoffderfizit der
   Gemeinschaftspapierindustrie ist eine der Hauptursachen der
   Schwierigkeiten dieser Industrie und einer der Gründer ihrer geringen
   Produktionskapazität, die durch immer bedeutender werden de Papier-
   und Pappeeinfuhren kompensiert werden muss.

b) Schaffung der Voraussetzungen für ein besseres Management der
   Industrie- und Haushaltsabfälle durch Sicherstellung einer
   wirtschaftlichen Wiederverwertung von Faserrohstoffen, die diese
   in grossen Mengenanteilen enthalten, wodurch gleichzeitig eine
   proportionelle Verringerung der zu beseitigenden Abfälle erzielt
   würde.

3. _Beschreibung des Programms_

In technischer Hinsicht ist festzustellen, dass es zur Wiederverwertung
zusätzlicher Mengen von Altpapieren un Altpappe erforderlich ist, alle
Qualitäten von Altpapier durch optimal geeignete neue Technologien und
Verfahren einer besseren Verwertung zuzuführen.

Zu diesem Zweck wurden vier Themenbereiche ausgewählt, die in ihrer
Gesamtheit den Hauptbereich des derzeitigen Forschungs- und
Entwicklungsbedarfs auf diesem Gebiet abdecken :

Forschungsthema   I. Charakterisierung der Aufbereiteten Fasern,
                      ihrer Veredelung durch verschiedene Verfahren
                      und die Wirkungen wiederholter Aufbereitung des
                      Fasermaterials.

Forschungsthema   II. Beseitigung der schädlichen Auswirkungen von
                      Kontaminations- faktoren im Altpapier.

Forschungsthema III. Entschwärzung und Behandlung der Abwässer von
                      Altpapieraufbereitungsbetrieben.

Forschungsthema   IV. Verwertung von Fasermaterial aus Haushaltsmüll,
                      sanitäre Probleme im Zusammenhang mit der
                      Verwertung solchen aufbereiteten Fasermaterials.

Raw material supplies to EEC countries have been causing serious concern for several years. Problems result because heavy dependence on the rest of the world stimulates balance of payments deficits and also gives rise to supply security risks.

Wood products with paper predominating, show the second largest trade deficit, immediately below that for petroleum products. Furthermore, the threat to the EEC paper industry from the highly competitive Scandinavian and North American industries is growing. With this in mind the Commission submitted a Communication to the Council concerning the problems posed by the paper industry in the Community (Document SEC (74) 1215 final). The Communication set forth the various possibilities for Community measures to safeguard the development of the paper sector.

Investigation revealed that :

the Community market is dangerously dependent on the outside world : net imports of paper materials represented more than 50 per cent of paper and board consumption in the last ten years ;

the enlarged Community's balance of trade in paper raw materials and products with non-Member States is heavily in deficit, and this deficit has increased substantially in recent years, despite world recession ;

the forestry and other vegetable sources in the Community supply only about one sixth of requirements and an increase in these resources can be envisaged to only a limited extent (doubling of current production) after a fairly long period (at least 20 years) following the implementation of an adequate and vigorous forestry policy ;

wastepaper has covered about 40 per cent of the Community paper and board industry's requirements of fibrous material in recent years, but this proportion represents only about 30 per cent of Community paper and board consumption.

Promoting more intensive recycling of paper and board is desirable because :

for a long time the paper industry has found recycling an effective way of offsetting supply problems and of withstanding competition from Scandinavia and North America ;

more recently public authorities have come to think of recycling as a
way of eliminating wastepaper from urban refuse and of reducing
dependence of the paper industry on the outside world.

Accordingly, in an initial report in November 1975, the CREST Subcommittee
on Raw Materials Research and Development recommended that a Working Party
on Paper should be set up to examine the situation in the Member States.
The Working Party was to determine what research work would be likely to
enable the level of recycling of paper-making materials to be raised and
the quality of recycled paper to be improved.

The Working Party which was set up in 1976, produced a set of suggested
R & D topics and projects. From these the proposal now being implemented
was prepared.

It should be borne in mind that the EEC environmental action programme
for the period 1977-81, adopted by the Council on 17 May 1977, calls for
a general policy of recovery, recycling and re-use, as part of the effort
to combat waste and of better management of refuse. One form of action to
be given priority is recovery and re-use of wastepaper.

The benefit to be gained from the research program are to be gauged from
the two viewpoints of raw materials supply, and environmental protection.

As a means of improving supplies of raw materials an improvement in
recycling techniques would make it possible for wastepaper to be used
where it is of little service at present, such as for printing paper and
paper for domestic uses and tissues. This would result in a drop in
consumption of both mechanical and chemical pulp.

The current rate of use of wastepaper in printing papers is currently very
much lower than its rate of use in other types, for instance paperboard,
corrugated paper and other types of wrapping paper.

At present recovered pulp can be used in the production of printing papers
in one of the following ways : either to replace chemical pulp or in
replacing mechanical pulp by incorporating old de-inked newspapers in the
manufacture of newsprint. (Federal Republic of Germany and the United
Kingdom).

In both cases the amounts available are limited and any increase in demand
upsets the price for these types of wastepaper, and is thus likely to
discourage their use.

More detailed research on recycling will reveal the possibilities of using further types of wastepaper. Application may include :

wastepaper containing wood (particularly magazines with a high percentage of wood) in order to obtain a pulp similar to mechanical pulp ;

and wood-free wastepaper (particularly leaflets) in order to obtain pulp of a type somewhere between mechanical pulp and chemical pulp from hardwood fibres.

If the characteristics of various recovered fibres are distinguised and this is combined with better removal of contraries, improvement in de-inking techniques and more intensive and advantageous recovery of urban fibres, it will be possible to make better use of many kinds of wastepaper. This will lead to more discriminating use of paper-making raw materials according to the needs of the products being manufactured.

Moreover, it will mean that some qualities of wastepaper now used as filler can be upgraded. For example, salvaged newspaper and leaflets now used in board-making could, with efficient de-inking, go into the manufacture of printing paper.

However, such a revolution is inconceivable unless board manufacturers can be offered an alternative raw material, such as urban fibres.

De-inked pulp from salvaged newspaper containing wood may become good enough to replace mechanical pulp in its traditional uses. This is important at a time when improved mechanical pulps (thermo-mechanical pulps) are ready for upgrading to higher quality products than those for which they have been used traditionally.

As regards de-inking, research is needed because present techniques are incapable of achieving satisfactory de-inking of many papers bearing offset or gravure printing. Moreover, the introduction of new kinds of ink (especially UV but also offset newspaper inks) is creating an urgent need for further research on de-inking. Research on the reduction of pollution caused by de-inking is also necessary.

This research should be carried out in collaboration with ink manu-facturers and with printers.

From the environmental point of view there are some obvious advantages in recycling wastepaper. For instance it reduces refuse, and saves wood. In addition making paper from wastepaper causes less water pollution than making it from wood.

However some recycling operations, especially de-inking, produce a certain amount of pollution but pollution deriving from de-inking is much less than that from the manufacture of chemical pulps, but it is nevertheless greater than that deriving from the manufacture of mechanical pulp or from a conventional treatment of wastepaper. Pollution is bound to be reduced by appropriate research.

On the basis of a survey of R & D in hand or planned on paper recycling and an assessment of current research needs in this sector, carried out with the CREST Working Party, the Commission considered it necessary to propose a set of measures which were taken up in the Council Decision of 17 April 1978.

Four main research topics were selected, namely :

characterization of reclaimed fibres, their upgrading by various processes, and the effects of multiple recycling on paper-making fibres.

elimination of the detrimental effects of contaminants in wastepaper, including the dispersion of thermo-softening contaminants.

de-inking, including the relationship between different types of ink and de-inking, and the treatment of effluent from wastepaper recycling plants.

use of urban fibres, including technological characterization of solid urban waste and health problems caused by the use of recycled fibres.

Each of these four topics was studied in detail and the following were defined :

framework of research to be undertaken on the topic concerned ;
general, technical and economic reasons for the research, together with the benefits expected from application of the results ;
time required for the research in question, with regard to existing resources (facilities, manpower) and short-term availability of competent research centres ;
estimation of the total cost of implementing the research programme.

The program is being implemented as an indirect action by means of
contracts, partly financed (usually in the proportion of 35 - 50 per cent)
from the Community budget, made with private or public research bodies in
the Member States.

Funding the program implies a maximum financial contribution of
2.9 million EUA from theEEC budget, of which about 300 000 EAU will be
used to cover the management costs of the program.

Offers to take part in the program comprised 37 research proposals,
consisting of :

    eight proposals for the first topic, of characterization of
    fibres :

    four proposals for the second topic, of elimination of the
    detrimental effects of contaminants ;

    fourteen proposals for the third topic, of de-inking ;

    nine proposals for the fourth topic, of use of urban fibres.

Two proposals involved subjects outside the frame of reference set for
the program. In addition to these two proposals, a further six were
discarded either because the research subject was covered by another
wider or better designed proposal or because the proposed research subject
was unsuitable for inclusion in this initial program.

It was possible to combine some of the other research proposals to avoid
dividing up the work too much and so that the research under the various
proposals could be more complementary.

After critical examination of the proposals and after discussions between
some of the applicants and the administrative units of the Commission,
26 research contracts were made. Their distribution among the four
chosen research topics is as follows :

    for the first topic, of characterization of fibres, seven contracts
    have been made, requiring financial assistance of about 700 000 EUA ;

    for the second topic, of elimination of the detrimental effects of
    contaminants, four contracts have been made, requiring financial
    assistance of about 625 000 EUA ;

    for the third topic, of de-inking, ten contracts have been made,
    requiring financial assistance of about 950 000 EUA ;

for the fourth topic, of use of urban fibres, five contracts have
been made, requiring financial assistance of 270 000 EUA.

Broadly speaking, the research undertaken under these contracts was to
take one or two years. Research began in 1979 and the first results are
expected at the end of 1980.

QUESTIONS

*In answer to a questioner pointing out that recycled paper costs more to
the consumer than conventional paper, Ing. Fassotte repeated that further
research was certainly needed into the problem of de-inking.*

RECYCLING SOLID URBAN WASTE

Dr Carlo Noto La Diega

Sorain Cecchini SpA

Italy

SYNOPSIS

Necessity of considering urban solid waste as a secondary raw material from which finished or semi-finished products can be recovered. Disposal of urban solid waste is a service that the community must provide. Consumers cannot afford any longer indiscriminate exploitation of the natural resources.

Hence birth of technologies aimed at recovery, recycling and reutilization. Distinction between destructive and preservative technologies : incineration and recycling. Examination of dry and wet system recycling.

Most interesting components of solid urban waste are paper, plastic, iron and organic matter. Description of Rome's recycling plants, the only ones in the world of industrial size with more than ten years actual work.

RESUME

Cet exposé se fonde sur la thèse que la nécessité, plus encore que la possibilité, de considérer les déchets solides urbains comme une matière première secondaire susceptible de fournir des produits finis ou demi-finis résulte de la prise de conscience d'une double réalité, à savoir que :

- *l'élimination des déchets solides urbains est un service que la collectivité est tenue d'assurer pour se conformer à la législation en vigueur et protéger l'environnement ;*
- *notre civilisation de consommateurs ne peut plus s'offrir le luxe d'exploiter sans discernement les resources naturelles.*

Ceci explique l'avènement de techniques de récupération, de recyclage et de réemploi dont les objectifs sont parfois différents et contradictoires. L'exposé établit une distinction entre techniques destructrices et techniques conservatrices parmi les premières on a l'incinération, parmi les secondes, le recyclage. Les systèmes de recyclage par voie sèche et par voie humide sont esquissés dans leurs grandes lignes.

Ils présentent un certain nombre de différences et peuvent fair appel à divers procédés suivant la composition des déchets, les produits à récupérer et les choix techniques fordamentaux. Le recyclage a pour but la fourniture de produits utilisables, à la place d'autres matériaux, dans les activités industrielles et agricoles courantes, sans aucun inconvénient d'ordre sanitaire, environnemental, technique ou commercial.

Il est notoire que dans les pays de la CEE les composants les plus intéressants des déchets solides urbains sont le papier, les matières plastiques, le fer et les matières organiques; c'est pourquoi l'orateur analyse la technique de tri envisagée, expérimentée ou mise en oeuvre dans les différents pays et se livre à un examen des domaines d'emploi appropriés notamment au contexte social et économique existant.

Il s'attache brièvement à décrire les installations de recyclage de Rome qui, d'après une étude détaillée de la CEE, sont les seules qui aient atteint la dimension industrielle, avec plus de dix années à leur actif.

Une série de 20 à 25 diapositives seront projetées pour illustrer les principaux aspects de ces installations.

Forschung in der Gemeinschaft

Aus den Argumenten in diesem Bericht geht hervor, dass die Notwendigkeit, weit mehr als die Zweckmässigkeit, feste Siedlungsabfälle als sekundären Rohstoff zu betrachten, aus dem Fertigprodukte oder Halbfertigwaren gewonnen werden können, auf die Anerkennung zweier Realitäten zurückzuführen ist:

- Die Beseitigung von festen Siedlungsabfällen ist eine Dienstleistung, die der Gemeinschaft in Erfüllung der Gesetze und zum Schutz der Umwelt obliegt;

- die Verbrauchergesellschaft kann sich die wahllose Ausbeutung
  der natürlichen Quellen nicht mehr leisten.

So kam es zur Entwicklung von Technologien zur Wiedergewinnung,
Aufbereitung und Verwertung mit zum Teil verschiedenen und sogar
einander widersprechenden Zielen. Der Bericht unterscheidet zwischen
zerstörerischen und erhaltenden Technologien: Zu ersterer gehört die
Verbrennung und zu letzterer die Aufbereitung. Der Bericht enthält
ferner eine eingehende Beschreibung der trockenen und nassen Aufbe-
reitungsverfahren.

Zwischen diesen Verfahren besteht eine Reihe von Unterschieden und
möglichen Prozessdiagrammen je nach der Zusammensetzung des Abfalls,
den wiederzugewinnenden Produkten und den grundlegenden technischen
Auswahlkriterien.

Zweck der Aufbereitung muss die Lieferung von Produkten sein, die
bei normalen industriellen und landwirtschaftlichen Tätigkeiten an-
stelle anderer Stoffe ohne Beeinträchtigung der Gesundheit, der Umwelt sowie der technologischen und kommerziellen Aspekte eingesetzt
werden können.

Die wichtigsten Bestandteile der festen Siedlungsabfälle in den
EWG-Ländern sind natürlich Papier, Kunststoff, Eisen und organische
Stoffe. So werden auch die in den einzelnen Ländern vorgeschlagenen,
erprobten oder durchgeführten Trennungstechnologien analysiert und
die einschlägigen Anwendungsbereiche auch im Verhältnis zu den vor-
herrschenden sozialen und wirtschaftlichen Bedingungen untersucht.

Ferner wird der Bericht eine zusammenfassende Beschreibung der Auf-
bereitungsanlagen in Rom enthalten, bei denen es sich nach einer ein-
gehenden Untersuchung der EWG um die einzigen grossindustriellen Anlagen in der Welt handelt, die nach mehr als zehn Jahren nunmehr er-
probt sind.

Zusätzlich wird eine Serie von 20 bis 25 Diapositiven gezeigt, um
die Hauptaspekte der römischen Anlagen darzustellen.

Universal rubbish collection services and the need to recover resources are two considerations that have prompted the development of various recovery methods. The differences between them arise from the fact that the composition of waste varies considerably from one country to another and because the success of recycling - in other words, the amount re-covered - depends on the socio-economic development of the country con-concerned.

The CREST final report on progress in recycling in the world shows that there are many pilot plants at the development stage. In Italy, the only industrial-scale plants which have been in operation for a number of years are in Rome and Perugia.

However in Rome or anywhere else, the first step towards recycling is to perfect methods of sorting out the various components.

Wet methods cover the technology for pulping refuse. After passing through "pulpers" the waste is sorted by exploiting the different behavioural characteristics of the suspended matter. For instance paper and board when pulped give fibres with a low density in suspension, while plastics, wood and many organic materials float and can therefore be separated by flo-tation.

Glass, earthenware fragments and metals are separated out by continuous centrifugation and ferrous metals are extracted with electro-magnets. Finally, the pulp, which is the inevitable end product, retains its fibrous matter until the paper refining plant removes it.

Dry sorting methods may be categorised according to the manner in which they separate waste. It may be pulverized or bags may be burst without changing the physical nature of the individual materials. After this stage, each method follows a different course depending on the type of waste. The type of plant and process may vary but the emphasis is always on the extraction of those materials present in sufficient quantity as to make recovery economically worthwhile, i.e. : paper and board, plastics, ferrous metals, organic matter, non-ferrous metals, glass, etc.

The method generally used for separating paper is a device called an "air classifier" which relies on the different specific weight of paper - compared with other materials - to separate out the lighter fractions from the waste stream.

Already several are in operation, or at the design stage. There are considerable differences between the various models and between actual or rated performance. Some merely separate out the lighter fractions, thus giving an end product composed not only of paper but of plastic film, rags and other light materials.

Other more sophisticated equipment achieves a very high concentration of paper and eliminates almost all extraneous matter. Machines of the first type are obviously designed to produce an end product which can be converted into fuel, while those of the second type are designed to recycle paper and board for reprocessing in the paper-mill.

There is a third method whereby the light fractions are separated out along with the inevitable extraneous matter and then other methods are used (mechanical, hydraulic or thermal) to produce the pulp for the paper-mill.

The choice whether paper is re-used after separation as an alternative fuel or as paper pulp depends on many considerations, but in Europe, the solution generally advocated is its use as pulp in paper-mills.

The technique for separating out plastic film is very similar to the one used for paper. The plastic can be re-used as a fuel, mixed with paper or used as plastic after further separation. This may be carried out hydraulically, in the pulpers; thermally, using hot air currents; or mechanically, exploiting the ways materials respond to being cut or broken.

After separation, plastic film may be re-used directly for moulding larger objects. The moulding plant is expensive and complex and the products obtained (pegs, inserts, pallets, bobbins, etc.) cannot always compete with those already on the market made from other materials.

It is far better to use the upgraded plastic to obtain a product in granular form which can either be moulded in normal industrial machines or be reprocessed into film. This is a more satisfactory method of recycling, especially as the cycle is repeatable.

FERROUS MATERIALS

Separation is very simple and high yielding when electromagnets are installed at the most strategic points.

Fragments separated have a very low specific weight and still carry labels, or contain various residues. They thus have to undergo further processing, including either using heat to burn off all the impurities prior to compacting the clean scrap, or milling to remove impurities and considerably increase its density.

The material obtained is used in the arc furnaces at any conventional steelworks. It is particularly welcome because it is "soft" and readily meltable.

ORGANIC MATTER

Methods of separating out organic substances are based almost exclusively on grading because they are all directed towards two final uses for this type of waste : dumping or composting.

Composting is more widely used when the total waste being treated contains more than 25 percent organic matter. After the composting cycle, the end product is cleaned, since it still contains the 25 or 30 percent of extraneous matter - glass, hard plastics, earthenware fragments, metals, wood, rags, etc. - which it had when it was graded. This cleaning operation is in fact rather difficult.

However we shall give attention to the use of the organic matter as feedstuffs.

For feedstuffs the organic matter cannot be separated out merely by sizing but has to undergo complete sterilisation in autoclaves. The end product is dry (8 percent humidity), in pellet form and has a protein value (12 - 14 percent) comparable with many conventional feedstuffs; it is also far more appetizing in view of its "noble" origin.

NON-FERROUS METALS

No great attention has ever been paid to separating out non-ferrous metals since they are usually present in waste in only very small proportions. The problem of aluminium can only really be tackled in those countries where it is present in significant quantities. Induced repellent currents may be used. A number of studies are being carried out into the extraction of all metals together by means of flotation. Other studies have concentrated on the electrolytic extraction of specific metals from incinerator slag.

Separation of glass is well worthwhile since quantities are considerable. There are two separation techniques available. "Froth flotation" is controlled flotation in special liquids, and "optical selection" is used on waste sized at between 6 and 20 mm with glass already present in large amounts. The waste passes a special optical cell and anything transparent is removed.

Plants in Rome treat 1 800 tonnes of solid urban waste per day, sorting it dry without any previous pulverization. After sorting, the materials are sent off to nearby processing and upgrading plants where they are made into the following end products :

(i)   pulp for paper-mills;

(ii)   plastic in granular form for moulding and film;

(iii)   baled scrap iron for steelworks;

(iv)   feedstuffs for farms;

(v)   agricultural compost;

(vi)   steam for use by the plant and for sale.

There is also a semi-industrial plant producing rdf (refuse-derived-fuel - patented by CALURB) which is either burned to produce steam and electricity or sold to the cement works at Guidonia.

FENCE SITTING BY INDUSTRY IS WASTEFUL IN THE LONG RUN

Dr Lisa Pavan

Environment and Consumer Protection Service
Commission of the European Communities

SYNOPSIS

The first generation of EEC ecological policies was designed to cope with a range of immediately pressing problems. The basic aim was to clean up pollution.

The change of orientation came in May 1977 with the second Action Programme on the Environment. "Prevention is better than cure" was its motto and since then the EEC Commission has been working at introducing new policy instruments specifically designed to improve environmental quality.

In Brussels we are not looking for the miracle solution which would at once and in all Member States minimise packaging manufacturing and distribution costs, maximise consumer satisfaction, reduce social costs and safeguard employment.

The Commission is however aware that the packaging industry throughout Europe could make a distinct contribution to environmental issues by designing products for re-use or multiple use, where suitable; ease of reclamation, recycling and disposal; using the least energy intensive material that will do the job, and reducing certain polluting effects of the manufacturing of packaging materials.

RESUME

La première génération des politiques écologiques de la Communauté euro-péenne avait été conçue pour faire face à toute une série de problèmes urgents, le but essentiel de ces politiques étant de lutter efficacement contre la pollution.

Le changement d'orientation est intervenu en mai 1977 avec le deuxième programme d'action en matière d'environnement. "Mieux vaut prévenir que guérir" fut son mot d'ordre et, depuis lors, la Commission travaille à l'introduction de nouveaux instruments de politique spécialement conçus pour améliorer la qualité de l'environnement.

Nous ne cherchons pas, à Bruxelles, de solution miracle permettant de
réduire immédiatement et dans tous les Etats membres les coûts de production et de distribution des emballages, de maximiser la satisfaction du
consommateur, de réduire les coûts sociaux et de sauvegarder l'emploi.

La Commission est toutefois consciente que l'industrie de l'emballage
en Europe pourrait, pour sa part, apporter une contribution véritable à
la solution des problèmes de l'environnement en concevant les produits
permettant une réutilisation ou des emplois multiples, selon le cas;
une récupération, un recyclage et une élimination faciles; l'utilisation
des matériaux à plus faible intensité d'énergie qui conviendraient; la
réduction de certains effets polluants des matériaux de fabrication et
d'emballage.

ZUSAMMENFASSUNG

Die erste Generation der politischen Massnahmen der EG auf dem Gebiet der
Ökologie war auf eine Reihe sofortiger dringender Probleme ausgerichtet.
Das grundlegende Ziel bestand in der Bekämpfung der Verschmutzung.

Eine Richtungsänderung kam in Mai 1977 mit dem zweiten Aktionsprogramm
für den Umweltschutz zustande. Der neue Leitsatz war: "Vorbeugen ist
besser also heilen"; die EG-Kommission hat sich seither um die Einführung
neuer politischer Instrumente zur Verbesserung der Umweltsqualität bemüht.

In Brüssel wurde keine Zauberlösung gesucht, die sofort und in allen
Mitgliedstaaten die Kosten der Herstellung und Verteilung von Verpackungen
auf ein Minimum herabsetzt den Verbraucher optimal zufriedenstellt die
Sozialkosten vermindert und die Beschäftigung sichert.

Die Kommission ist sich folgender Tatsachen bewusst die Verpackungsindus-
trie in ganz Europa könnte einen wertvollen Beitrag zur Lösung der
Umweltprobleme leisten, wenn sie Erzeugnisse unter Berücksichtigung
folgende Anforderungen entwickeln würde: ggt. Wiederverwendung oder
Mehrzweckverwendung; Erleichterung der Rückgewinnung, Wiederverwertung
und Beseitigung; Verwendung des am wenigsten energieaufwendigen Materials,
das für den betreffenden Zweck geeignet ist; Verminderung bestimmter
Verschmutzungswirkungen bei der Herstellung von Verpackungsmaterial.

The environment has now established itself pretty well in the language of politics. Quite quickly too.

The U.K. Department of the Environment is not yet ten years old. In the 1970s, the country which is today hosting this Conference acquired a permanent Royal Commission on Environmental Pollution and a Commission on Energy and the Environment. The Department of Education and Science has had a report on Environmental Education in the U.K. The catalogue could be extended.

We are now moving on from the environment to ecology. According to Lord Beaumont in a House of Lords debate last year, what we are now talking about is "the ecological perspective". I will only mention in passing the recently animated and very lively Ecology Party, the newly formed Green Alliance and other campaigning environmental and ecological pressure groups in England and "on the continent".

What, though, is ecology ? As a branch of science, it is defined as the study of the relationships between living organisms and between organisms and their environment, especially animal and plant communities, their energy flows and their interactions with their surroundings. Barbara Ward has a good phrase which puts it in its political context : "rules for the care and maintenance of a small planet".

The idea that mankind is just one species in an incredibly complex web of interacting and interdependent life is not new.

Some 120 years ago in the Northwest territories of the United States, the Suquamish Indian Tribe faced near extinction in its confrontation with the westward movement of white society. In his final negotiation with government agents forcing the Suquamish onto a reservation, Seattle, chief of the tribe said some notable things that lay ignored as curious cultural irrelevancies for over a hundred years. He said : "... the earth does not belong to the man ; man belongs to the earth .... all things are connected .... Man did not weave the web of life ; he is merely a strand in it. Whatever he does to the web he does to himself".

Over a century has gone by. Today, we still do not fully realize the complexity of the interactions between men and the environment. We still do not recognise the self-destructiveness of man's technological and industrial power.

Once this recognition takes hold of politics, it will transform all the conventional terms of reference. It is urgent that it should, if we want to avoid the disruption of the ecosystem in which we all live.

An illuminating case history is the European Agricultural Policy. The CAP resulted from strictly economic and political bargaining and aimed to achieve purely economic and political goals.

During the last fifteen years we have witnessed an important increase in productivity. But, at the same time, intensive farming, the elimination of Wetlands, the substitution of pastures by arable, a complete lack in land-use planning, have weakened the social as well as the ecological fabric of our countryside.

In the Commission we believe that we have now embarked on a determined effort to ensure that any new model CAP adequately reflects the environmental viewpoint. Perhaps it is a sign of advance that to many of you this will now sound common-place. But it is not yet the common-place of politics which it should be.

I mention this to make the point that the delay in coming to terms with the ecological imperative, here and in the other European countries is a particularly serious one.

In some situations if you delay a decision long enough the problem answers itself. This in a way is true of ecological constraints, but only in a catastrophic sense. Bad farming corrects itself by creating deserts, eliminating man the exploiter and thus bringing about a new balance.

It is nowadays hardly possible to carry out a harmonious development of economic activities and a continuous and balanced expansion - which constitute the paramount aims of the European Community - without protecting natural resources, increasingly important economic and social factors, from being damaged in a way which might be irreversible.

Until recently we have looked at the problem of waste from the point of view of eliminating pollution. The EEC Council has adopted a framework directive on waste management and specific directives on the disposal of toxic and hazardous waste, waste oils, titanium dyoxide and other particularly noxious chemicals. But we see now that this is only tackling a part of the problem and we are trying to add a new dimension to our work in this area.

At the heart of the 2nd Action Programme of the Community on the Environment, which was adopted by the Council of Ministers in May 1977, is the idea that a sound environmental policy must include the concept, in the broadest sense, of resource management.

The availability of energy and of natural resources in general, in a world where both the population and its material demands are constantly growing, is a major long-term reality which we all have to face. The challenge is to see how short-term political decisions can be properly framed to face these and other long-term realities in an effective and humane manner.

With regard to waste therefore, we have to spend more time not only trying to reduce it at source but also devising means to encourage its re-use, recycling and recovery. And there are not only good environmental reasons for so doing. The economics of the operation are also quite interesting.

If we could extract from the waste stream the potentially recoverable materials that are thrown away, we could achieve a saving of some 5 billion pounds each year over and above the cost of recovery.

Not surprisingly therefore, it is the rising price of raw materials and dependence on imports that has provided the major impetus to take measures to reduce waste both at national and at Community level.

The objectives of sound resource management have been clearly stated, and we can now say that it should be the policy of the Community as a whole and of its Member Nations to :

1. provide adequate energy and materials supplies to satisfy
    not only the basic needs of nutrition, shelter and health
    but also those of a dynamic economy without indulgence in
    waste ;

2. rely on market forces as a prime determinant of the mix
    of imports and domestic production in the field of materials
    but at the same time decrease and prevent, wherever necessary,
    dangerous or costly dependence on imports ;

3. accomplish the foregoing objectives while protecting or
    enhancing the environment in which we live ;

4. conserve our natural resources and environment by treating waste as materials and returning them either to use or, in a harmless conditions, to the ecosystems; and

5. institute coordinated resource policy planning which recognizes the interrelationships among materials, energy and the environment.

Why, you may ask, judging from the Commission's areas of interest, does urban waste, seem to be considered one of the worst "offenders"?

(i) The immediate reason quite simply is an economic one. Collection and disposal costs are becoming prohibitive. As such they represent an increasingly heavier burden for local authorities and the payer of rates and taxes.

(ii) But we also have to put the problem into its ecological perspective. We now realise that the energy crisis, the raw material crisis, the natural resource crisis (whatever you wish to call it) is not a temporary phenomenon. Its underlying causes are deeply rooted in the patterns of production and consumption which have over the last several decades been developed in the industrialized world.

To strike a balance between the "need to produce" and the "need to protect", it is important that rational management and preservation of natural resources be encouraged.

Domestic refuse is a field which lends itself beautifully to achieving two complementary goals at once. It offers a large scope for:

- reducing waste and preventing nuisances, on the one hand
- and preserving natural resources on the other, thanks to further recovery of raw materials and the energy which they produce.

(iii) Last, but not least, it is really the public authorities who are responsible, either directly or indirectly, for the disposal of urban waste. In this connection, I think it is worth mentioning that until recently the environmental damage costs caused by household refuse have mainly been borne by society in general or "third parties", rather than the specific industries and customers whose decisions caused the damages.

Waste generators and disposers have seldom been charged directly for refuse disposal services the way they have been for other public utilities. I will not go into details as to what caused such a situation, but I do wish to underline that, as a result, waste management facilities have been underestimated by their users; and consequently market forces have failed to develop fully the necessary economic incentives to reduce waste and stimulate the productive use of secondary materials.

In conserving materials through accelerated urban waste recycling and greater efficiency-of-use of materials, it is important that environmental costs be taken into account. At the same time, all resources, including environmental, must be paid for by users.

Before Autumn 1973 little attention was paid to the conservation of energy and other natural resources which could be obtained through an economically viable and ecologically sound management of urban waste. It is now becoming clear, however, that the question of future energy and raw material availability will not be solved by locating a single massive source of savings - significant conservation depends on the accumulation of many modest efforts at waste reduction.

The fundamental need therefore is to develop and promote alternative resource strategies. By alternative resource strategies I mean strategies which

(i)   are the most economical in the use of non-renewable materials;

(ii)   have the smallest impact on balance of payments; and

(iii)   are least harmful to the environment.

At the simplest level this means promoting research and development of alternative sources of supply: solar energy, nuclear fusion, the recovery, re-use and recycling of every kind of energy and materials - all these will be of importance.

More fundamentally, however, the need is to act on the demand side of the equation; to reduce the rate at which demand for energy and materials is growing and perhaps, ultimately, to reduce even the absolute level itself.

This came out quite clearly, I think, during two recent meetings of the Council of Ministers. In December 1978 and April of last year the Council held an exchange of views on the tightening up of the Community's environmental policy.

When discussing waste management, the Ministers covered four specific sectors, namely: waste paper, old tyres, waste oils and beverage containers. They examined, for each in turn, the solutions which take the interest of society as a whole best into account.

Each of these sectors represents only a small part of the total Community flow of solid waste. Nevertheless, they all entail substantial and increasing social costs.

Take the single case of packaging. It accounts for 20 per cent to 30 per cent by weight and 50 per cent by volume of urban refuse. The manu-facture of packaging demands considerable expenditure of energy and other natural resources; it contributes to air and water pollution. Used packages may become litter or end up in a dustbin. In both cases they add to the growing mountains of municipal refuse.

We are again talking about a relatively small percentage of the solid waste generated in the nine Member States - something in the order of 1.5 per cent of the total tonnage. Similarly the energy and other natural resources used to manufacture, distribute and dispose of packaging are not a major part of global material consumption, and, in addition, packages are designed to conserve and give adequate protection to their contents. They obviously contribute to the avoidance of waste. Think for example about perishable foodstuffs: a case which speaks for itself.

In absolute terms we are nevertheless talking about amounts of energy and materials worth taking into consideration, and, as in the case of tyres, paper and waste oils, social costs entailed by air and water pollution, waste disposal, litter, and so on are not borne by those who use them, i.e., manufacturers, distributors and consumers.

Take the case of beverage containers. Even a rough estimate is hard to make but it shows social costs to be substantial (some 1400 million pounds in the nine Member States).

Social costs vary considerably with the type of container: they are much higher for non-recycled, one-trip glass than for plastic, and normally higher for plastic than for returnable glass.

Obviously packaging manufacturers have adapted to the rising costs of materials and energy. Designers have produced new containers, used other types of materials, but since external costs are not included in the price of packaging, the market has not caused the best distribution between the various forms of containers to be achieved.

This is one of the reasons why the EEC Commission's Environment Service has been working in this field and the Council of Ministers has considered various ways of

    (i)   re-using and recovering beverage containers, and

    (ii) improving the recycling of packaging materials.

More generally, I think we must look towards a reorganization of the patterns of production. Market deficiencies should, to my mind, be mitigated by changing and, if necessary, limiting the production, consumption and discharge of containers. This could be achieved by making lighter products, by recycling discarded packaging materials or by changing the way in which containers are used - an obvious example is that of a switch to returnable or refillable containers.

In Brussels we are not looking for the miracle solution which would at once and in all Member States minimise drink container   manufacturing and distribution costs, maximise consumer satisfaction, reduce social costs and safeguard employment.

The Commission is however aware that:

    1.   Certain measures taken by Member States in isolation may hinder trade and conflict with Community law, and

    2.   A common approach to reduction of urban waste and the conservation of raw materials and energy is desirable.

As far as beverage containers are concerned this common approach points in two complementary courses of action; i.e. the re-use of bottles and the recycling of materials.

European industry could make a distinct contribution to environmental issues by designing products for:

    1.   Re-use or multiple use, where suitable.

    2.   Ease of reclamation, recycling and disposal.

    3.   Using the least energy intensive material that will do the job.

    4.   Reducing certain polluting effects of the manufacturing of

packaging materials.

How did we reach these conclusions? In the work carried out so far by the Commission, two types of analysis have been conducted. We have tried, first of all, to identify the positive and negative aspects and effects of the various policies which could be envisaged.

This qualitative analysis was then followed by an examination of the socio-economic and environmental impacts deriving from the implementation of the various regulating policies. In this second phase, we tried to collect hard data on the drink container markets under examination (i.e. production, distribution and consumption figures). The energy, labour, pollution, raw materials impacts were then worked out with the help of a forecasting computer model.

What are the main findings of these reports? We found out that it is difficult to make everybody happy. To quantify in planning terms a solution which would satisfy packaging manufacturers, the drink industry, retailers, consumers and local authorities would be an almost impossible task.

The difficulty does not however preclude confirmation of what would be a desirable line of conduct on the part of the public authorities, namely that it would be to the benefit of all member nations to limit the amount of bottles, cans and cartons produced, consumed and disposed of in Europe.

Both the qualitative and quantitative analyses carried out confirmed furthermore that the three most appealing approaches to solve the problems posed by drink packages are:

(i)   introduction of a mandatory deposit scheme on glass containers
(ii)   to levy a tax on the production of all types of beverage packaging
(iii) to encourage recycling.

Conclusions

One of the basic principles in resource management involves, in my opinion, communication and candour. Environmental problems must be promptly recognised and widely discussed as soon as they are perceived.

A satisfactory solution can only be obtained when all parties communicate openly, honestlyand freely from the beginning and recognise that the ultimate solution may well be a compromise.

The kind of communication that has often typified resource management issues in the past as well as today is a sort of communication between adversaries.

I believe that we should change this. In Brussels we have been trying to hear all interested parties - industry, consumers, environmentalists - and give serious consideration to the concern of each of them.

I am convinced that we are now moving in the proper direction and we need everybody's - and particularly industry's - co-operation and support.

We need it, and I think the issue deserves it, because if we all concentrate our attention on problems that confront us today, while giving little attention to those we will face tomorrow, we can all sit safely on our fence without worrying too much about the state of the environment inherited by future generations.

PACKAGING - THE INTER-PROFESSIONAL POSITION

Dr Werner Hoffmann

European Container Glass Federation

SYNOPSIS

Figures and Data on Packaging
Addendum and interpretations.

Development of Packaging to its Present Scope
- *Costs of labour vs. costs of industrial products.*
- *Rising living standards and changing consumers' habits.*
- *Consequences : changing modes of production and distribution re-quiring new forms of packaging as demonstrated e.g. for the market of liquid foodstuffs and beverages.*
- *The complexity of interdependencies, the elasticity of the relevant markets and their susceptibility to eternal interventions.*
- *The real and alleged impacts of packaging on environmental parameters and the position taken by public authorities to account for the situation.*
- *The necessity of weighing ecological benefits against economic distortions caused by possible interventions.*

Possible Means of Intervention and their potential Consequences
- *Legal or executive intervention and their instruments; critical appraisal.*
- *Voluntary agreements between the Professions and public authorities, their instruments and critical appraisal.*

Actions Undertaken, Programs in Operation
Example glass and can recycling, results.

Optimal Solutions to be Expected
- *The immanent efficiency of market forces to achieve optimal results,*

both ecologically and economically, within a frame-work of rules.
- The role of the public authorites in this context.
- The instruments to be employed.

Perspectives
New partial waste separation systems as a basis for new forms of intensive recycling.

<u>RESUME</u>

Données chifrées et autres sur les emballages
Addendum et interprétations.

Evolution jusqu'à ce jour de la situation en matière d'emballage
- Coût de la main-d'oeuvre/coût des produits industriels.
- Augmentation du niveau de vie et changement dans les habitudes des consommateurs.
- Conséquences : changement des methodes de production et de distribution, d'où la nécessité d'avoir de nouveaux types d'emballages (cf., par exemple, le marché des aliments liquides et des boissons).
- Situation complexe de l'interdépendance, élasticité des marchés en cause et leur sensibilité aux interventions extérieures.
- Incidence prétendue des emballages sur les paramètres en matière d'environnement et incidence réelle; attitude des autorités publiques à cet égard.
- Nécessité de comparer les avantages sur le plan écologique avec les distorsions économiques dues à d'éventuelles interventions.

Moyens d'intervention et leurs conséquences éventuelles
- Législation ou interventions réglementaires et leurs instruments; évaluation critique.
- Accords volontaires entre les professions et les autorités publiques; instruments et évaluation critique.

Actions entreprises, programmes en cours
Exemple : recyclage du verre et des conteneurs métalliques, resultats.
Solutions optimales prévisibles
- La tendance naturelle des forces du marché à obtenir des résultats optimaux tant sur le plan écologique qu'économique dans le cadre des régles en vigueur.

- *Le rôle des autorités publiques à cet égard.*
- *Les instruments à utiliser.*

Perspectives

Nouveaux système de separation partielle des déchets annonçant l'apparition de nouvelles formes de recyclage intensif.

ZUSAMMENFASSUNG

Zahlen und Daten über Verpackung
Nachtrag und Interpretationen.

Entwicklung der Verpackung zu ihrem gegenwärtigen Umfang
- Gegenüberstellung von Personalkosten und Kosten der industriellen Erzeugnisse; steigender Lebensstandard und veränderte Verbrauchergewohnheiten.
- Folgen : Veränderte Produktion und Verteilung, die neue Formen der Verpackung erfordern, wie z.B. für den Markt von flüssingen Nahrungsmitteln und Getränken aufgezeigt.
- die Verwicklung der gegenseitigen Abhän gigkeiten, die Elastizität der jeweiligen Märkte und ihre Empfindlichkeit gegen Eingriffe von aussen.
- die tatsächlichen und angeblichen Auswirkungen der Verpackung auf Umweltparameter und die Haltung der öffentlichen Behörden zu dieser Situation.
- die Notwendigkeit einer Abwängung des ökologischen Nutzens gegenüber wirtschaftlichen Verzerrungen durch etwaige Massnahmen.

Mögliche Mittel des Eingreifens und ihre Folgen
- Gesetzlich e oder ministerielle Interventionen und ihre Instrumente; kritische Bewertung.
- freiwillige Abmachungen zwischen den Berufszweigen und den öffentlichen Behörden, ihre Instrumente und kritische Bewertung.

Ergriffene Massnahmen, Laufen de Programme
Beispiel Altglas- und Konservenverwertung, Ergebnisse.

Zu erwartende optimale Losungen

- *die Wirksamkeit der Marktkräfte zur Erzielung sowohl ökologisch als auch wirtschaftlich optimaler Ergebnisse innerhalb eines Rahmens von Regeln.*
- *die Rolle der öffentlichen Behörden in diesem Zusammenhang.*
- *die anzuwendenden Instrumente.*

Perspektiven

Neue teilweise Abfalltrennungssysteme als Grundlage für neue Formen der intensiven Verwertung.

Discussion on drinks packaging tends to centre on the one-sided contrast between throw-away and returnable containers and is beset by emotion that results in an overestimate of the ecological importance of this packaging.

This can be corrected by a general view. While it is usually regarded from only the aspect of potential waste and consumption of materials and energy, its function in a modern distributive economy is forgotten.

One very important factor which has favoured the development of packaging to its present state is the much faster rise in price of services than finished industrial products. Thus the competitive advantage is given to supplying prepacked goods against the system of measuring out quantities from the seller's bag. Wherever a one-way container costs less than returning, transporting and cleaning a returnable container, it will be given preference.

The rationalizing effect of packaging, especially disposable packaging has undoubtedly contributed to the fact that retail food prices have increased less rapidly in recent years than the general cost of living.

Every intervention which does not allow for the multiple interactions and capacity of modern distribution channels will just as surely result in price increases or in a dislocation or shrinkage of industrial activity. This means a deterioration in service to consumers, reduction in demand and finally a loss of jobs, all without any assured ecological advantage.

Whether these consequences should be permitted is a political question, but there is no point in trying to complicate the problem by arguing alleged reduced "social costs" or additional jobs created by such intervention.

"Social costs" is a term used to denote the burden on the environment
through air pollution, waste disposal, and the results of a possible
shortage of energy and raw materials which, it is said, are not allowed
for in market prices and are assumed to be higher for disposable than for
returnable packaging.

This attitude is only partly correct. A re-usable container must be more
durable, i·e·heavier than a disposable package, and therefore uses more
energy and materials. The return transport for refilling and the
cleaning of containers uses additional energy and creates effluent

problems which do not arise with disposable packages. Consequently, a
re-usable container must circulate many times to make up for these
additional "social costs".

With regard to emission nuisance, the expenditure by industry in
accordance with progressive clean-air legislation is reflected in costs
and therefore in the product price. The remaining "social costs",
ie unavoidable effects on the balance of nature or long-term shortages of
raw materials, cannot be measured, except insofar as the market has
already taken account of the shortage by raising prices, and therefore
cannot be used as a standard in judging the individual measures under
discussion.

Hopes of creating additional jobs through the change-over from disposable
to returnable containers miss the point that, as a result, skilled jobs
will be replaced by unskilled and that the loss of skilled jobs will
result in the shutdown of valuable production facilities until the
existence of whole sectors is threatened.

Furthermore, there can be no question of a mere shift in jobs if, for
commercial or practical reasons, the market cannot be fully supplied by
returnable containers and consumption is restricted as a result. The
effect would be shrinkage in an entire sector, undoubtedly affecting the
whole economy.

All proposed measures should, therefore, be judged according to whether
they bring maximum ecological advantages and avoid or reduce economic
disadvantages. Measures must also be in relation to the intended
advantage; this is undoubtedly not the case if, for example, restrictive

measures against one packaging material are extended to all potential substitutes, owing to the law against discrimination. Furthermore, regimentation of one sector of a free market economy will undoubtedly affect the entire economic structure.

First of, the proposed individual measures is the prohibition of throw-away containers. This still seems to be popular and has already been enforced in individual sectors in some countries. From the legislative viewpoint, prohibition is undoubtedly the simplest but also the worst method, since the economic effects are the most drastic. It is also a measure incompatible with a free economy.

Returnable systems require a minimum turnover, since otherwise full use will not be made (eg in trade) of the staff and resources available for handling return of deposits and storage.

Return systems are also transport-intensive and therefore highly dependent on the use of empty transport facilities, which must meet reauirements in respect of the time and volume of deliveries. As a rule, these conditions can be taken into account and satisfied only in regional distribution systems with rapid turn-round.

If, as a result of prohibition of disposable containers, there is a forced transition to returnable containers where these conditions cannot be fulfilled, the result will always be a price increase to the consumer, possibly associated with a reduction in supply until there is a shortage.

On the other hand the ecological success of prohibition is questionable. Re-usable containers are not ecologically preferable to disposable containers if they circulate only a few times. However, circulation is likely to be reduced wherever the operating conditions are unsuitable for the return system. It therefore appears that prohibition does not achieve a balance between ecological advantage and economic disadvantage and involves a disproportionately drastic intervention in the economy.

The mandatory deposit has scarcely less drastic effects, either as a single measure or in support of a prohibition on disposable containers. In order to understand the situation, the function of a deposit must be discussed briefly.

In a return system the deposit is intended to ensure the stock of bottles for repeated filling. Consequently the full value of the container as

packaging for the next cycle, instead of a new package, justifies the
expense of repaying the deposit, return transport and cleaning.   In a
return system of this kind, the bottlers' own interest usually ensures
a regular repayment of deposits.

Where this is not the case, it is preferable not to charge deposits by
legislation but by licensing regional deposit cartels, as has been done
successfully, for example, in the case of beer-bottles in the Federal
Republic of Germany.   In a contractual agreement of this kind, better
account can be taken of local conditions regarding return and refunds with
a view to accelerating the circulation of bottles than by the inevitably
rigid regulations in an inter-regional act.

The deposit, however, becomes an absurdity if it is also demanded for
disposable containers, to enforce their return for re-use as raw material.
In this case the deposit does not secure the useful value of the package
but only the minimum scrap value, which commercially does not cover the
cost of return, repayment of deposit and storage at the place of return.

In the Federal Republic of Germany, these costs are estimated at between
DM 0.06 and 0.12 per package, compared with the scrap value of only
DM 0.01 for a 200 g glass bottle.   This proportion will be similar
in other countries and scarcely any more favourable for containers made
of other materials.

In addition, almost insoluble problems of space and hygiene arise
throughout the distribution channel, particularly in trade, if containers
intended only for recycling have to be returned and disposed of, a task
for which traders have neither the space nor the capacity.   The result
of this kind of compulsory deposit, ie compulsory recycling, can clearly
be seen from the development of the drinks market in the state of
Michigan since the introduction of compulsory deposits.

Admittedly there has been a marked reduction in the use of disposable
containers, but this is offset by considerable price rises, a reduction
in supply and a marked decrease in consumption.   Clearly, mandatory
deposits of this kind will be no more useful than prohibitions in reaching
a reasonable balance between ecological advantage and economic
disadvantage.

The last group of government interventions is taxes or duties;   these can be levied on all containers or just on disposable containers. This method is discussed mainly from the aspect of financing the payment of "social costs". Since these taxes or duties do not directly affect distribution, they avoid the direct, drastic consequences of prohibition or obligatory deposits.

If the duties are low they remain ineffective, assuming that the consumer will accept the resulting price increases without changing his consumption habits. If they are high, they will favour the profitability of returnable over disposable systems and result in a considerable increase in the use of returnable containers.

The overall effect, however, will be that the market will be supplied at a higher price with some dislocation of industrial activity to the disadvantage of the packaging industry. Finally, if the tax or duty is so high as to prevent economic use of one-way containers, it will be equivalent to a prohibition, with all the previously-mentioned consequences.

These complex problems are difficult or impossible to solve by rigid, inevitably generalized legal regulations, and they cannot be settled by state institutions, which are not adapted to such tasks. This does not mean, however, that state activity is excluded from this sector.

On the contrary, the function of the state, in cooperation with the economy, is to set outline conditions with realistic objectives. This is necessary, in my view, because individual undertakings in a competitive economy cannot alone be expected to take the initiative voluntarily, unless they can expect that competitors will obey the same conditions. The state can also give support, eg by setting aside waste-disposal laws against organized recycling, or by promoting research and helping to educate the public. In other respects, it should be left to free market forces to find commercial solutions for achieving the objectives in question.

One good example is the West German waste disposal programme for 1975, together with the recent agreement between the French Government and associations of the packaging industry, bottlers and the trade, according to which the sector involved is obliged to reduce energy and waste by recycling, weight-saving and other measures.

Industry, in cooperation with consumers, is able to find solutions bringing considerable economic advantages. A good example is the success of waste glass recycling, the first organized action of its kind for reducing environmental pollution. In Switzerland, for example, glass recovered from households already amounts to 37 per cent of the total production of glass for containers. Corresponding efforts by the can industry show similar success.

Recycling will be particularly important in the future since it can attain the objectives of waste disposal, ie saving of resources and energy and avoidance of waste without or with only slight economic dislocation. Organized recycling is only in its infancy and is capable of considerable development. In this connection the following preconditions appear important, and apply not only to the recycling of packaging materials.

Firstly, production techniques must be adapted to the increasing use of secondary raw materials obtained by recycling, thus providing sales outlets for these materials, without of course reducing the safety or quality requirements for a particular use. On the other hand, improved methods of separation and processing must be developed, with particular emphasis on new ways of collecting these secondary raw materials which are not only cheaper but also reduce the facilities required for processing.

It must be remembered that all these efforts depend on cooperation by consumers, ie the citizen, who must not be over-loaded.

He cannot in the long term be expected to separate a number of useful materials in his house and bring them to a collecting-point. However, he may be expected to deposit certain materials, eg glass, cans and paper, in a particular section of his dustbin, which can be emptied in the same operation as other domestic rubbish.

In a system of this kind, which is now being successfully tried in Constance, some of the useful material in domestic refuse can easily be separated for further processing, thus obviating the need for expensive separation devices and methods. Of course this system implies re-thinking of the total organization of waste disposal and it may not be applicable everywhere, but it does seem worth mentioning as one of many possible solutions.

In the search for optimum solutions, state intervention, which will ultimately be expensive to the consumer, is not the best way. The objectives can be more easily achieved by using the forces of a free economy.

The chairman, Mrs Aloisi de Larderel, described the subject of containers as something of a sensitive and burning issue.

Dr Pavan commented that, for beer, the best option to encourage returnables was a combination of the deposit system and a tax on beer sold in non-returnable containers. The tax should be set at 10 per cent of the retail price.

The result of these measures would be a near halving of the content of beer containers in municipal refuse and a reduced energy consumption in the beer container industry by 10-12 per cent. It would also give benefits in the form of increased employment and reduced raw material consumption.

PACKAGING

Michel de Grave

Social and Economic Committee

of the European Communities

SYNOPSIS

The wasteful society in which disposable objects are produced has many
consequences for consumers. Consumers pay for the packaging which
some producers still call "free packaging", sometimes accounting for 15,
20 or even 30 per cent of the retail price; they pay again as ratepayers
and taxpayers for this packaging to be cleared away. They have to put up
with pollution of the environment by either litter or collected refuse.
Most important of all, they live in a kind of society that sacrifices the
future to the present and sensible ways of managing things to the
convenience and quick profit of some of its members.

The objectives to be attained might be identified as follows :

1. reduction of the weight of containers ;
2. recycling waste material and arranging selective collection ;
3. reverting to some extent to returnable containers, and standardizing
    bottles ;
4. using packaging materials which cause less pollution wherever possible ;
5. not concentrating on packaging alone but seeking consumer goods that
    last or can be repaired.

RESUME

La civilisation du gaspillage et de l'objet à jeter a de multiples
conséquences sur le consommateur. Il paie des emballages que certains
producteurs appellent encore "emballage gratuit" et qui représente

parfois 15 pour cent, 20 pour cent voire 30 pour cent du prix de vente au détail; il paie une deuxième fois comme contribuable pour l'élimination de ces emballages. Il voit son environnement pollué par les décharges sauvages ou non. Enfin, et surtout, il subit un type de société qui sacrifie l'avenir au présent, la solution rationnelle à la facilité et aux bénéfices immédiats de certains.

Les objectifs à atteindre pourraient être définis comme suit :

1. Réduire le poids des emballages.
2. Recycler les déchets et développer la collecte sélective.
3. Revenir partiellement à l'emballage consigné, et standardiser les bouteilles.
4. Revenir chaque fois que c'est possible à des matériaux d'emballage moins polluants.
5. Ne pas voir que les emballages. Promouvoir la durabilité et les possibilités de réparation des biens de consommation.

Zusammenfassung

Die Verschwendungs- und Wegwerfgesellschaft hat zahlreiche Folgen für den Verbraucher. Er zahlt für Verpackungen, die von einigen Herstellern noch als "kostenlose Verpackung" bezeichnet werden und manchmal 15, 20 oder sogar 30% des Einzelhandelspreises ausmachen. Er zahlt ein zweites Mal als Steuerzahler für die Beseitigung dieser Verpackungen. Er erlebt die Verschmutzung der Umwelt durch nicht überwachte oder überwachte Müllabladungen. Schliesslich und insbesondere erträgt er eine Art von Gesellschaft, die die Zukunft der Gegenwart und eine rationelle Lösung der Bequemlichkeit und den unmittelbaren Gewinnen einiger weniger opfert.

Die zu erreichenden Ziele könnten wie folgt definiert werden:

1. Verringerung des Verpackungsgewichts

2. Wiederverwertung der Abfälle und Zunahme der Spezialsammlungen

3. Teilweises Zurückgreifen auf berechnete Verpackungen
   und Normung der Flaschen

4. Soweit wie möglich Verwendung umweltfreundlicher Verpackungsmaterialien

5. Vermeidung einer ausschliesslichen Konzentrierung auf Verpackungen;
   Förderung der Haltbarkeit und Möglichkeiten des Ersatzes für Ver-
   brauchsgüter

In its preliminary programme for consumer protection and information policy, adopted by the Council more than five years ago, the Commission stressed that : "The consumer should no longer be regarded merely as a purchaser and user of goods and services for personnel, family or group purposes but as a person concerned with the various aspects of social life which may affect him directly or indirectly as a consumer".

In point 30 of that programme the Commission specified that : "In order that the individual and collective needs of consumers may be better satisfied, an effort should be made to seek solutions to certain problems of a general nature.

These problems include the prevention of waste (especially in the packaging of products), the useful life of goods and the problem of recycling of materials.

As for consumer movements, they are now no longer just content with producing product comparison studies, and other measures to protect consumers, but feel that the time has come to move on to the concept of the promotion of consumer interests overall. In other words consideration must be given to gearing the economy to consumer needs taking also into account energy policy as well as the environment.

The environment may be defined as "simply a special consumer facility providing benefits free of charge". Consumer organizations are increasingly aware of such matters and are therefore troubled that the Commission assigns them only a token association with its work.

The wasteful society with the production of disposable objects has many consequences for consumers. They have to pay for packaging, which can account for 15 to 30 per cent of the retail price and they pay again as taxpayers for the packaging to be cleared away. Then again they have to put up with pollution in the form of litter or collected refuse. Most important of all, they live in a society that sacrifices the future to the present and to a quick profit for some.

Truly, one wonders where the resources would come from if the developing countries took to using and wasting as much plastic and paper packaging as we do.

One unfortunate feature of the free market economy is that it puts short-term profitability before long term considerations. It is therefore impossible to achieve certain desirable aims without interfering with some free market prerogatives.

The European Economic Community produces 1 800 million tonnes of refuse per year, or 5 million tonnes a day. The potential value of all this unrecovered material is probably more than Bfr. 400 000 million a year and savings on imports of between Bfr. 200 000 and 350 000 million are estimated to be achieved by recycling it.

For 1977 it was estimated that local authorities had reached 100 000 million (£ 1 500 million sterling) in expenditure on collecting house-hold refuse alone.

Since the war there has been a great increase in pre-packaging with its advantage, but also resulting in a much bigger volume of waste materials and higher costs of refuse treatment. We are told that it costs more to salvage packaging materials than to throw them away and make new ones. Of course pollution is cheaper when those responsible for causing it are not made to pay the expenses it creates.

But even if disposable packaging does overall cost less, its use is nevertheless short-sighted, because considerations other than the financial ones have to be taken into account. The Commission itself has emphasised the importance of the sparing use of "resources in short supply" so far as petroleum-based products and metals are concerned.

Apart from waste involved with packaging, waste in all forms is to be condemned. One culprit is the "disposable" object, the device so designed that it is not intended to be repaired.

The problems are enormous. But what has been done so far to solve them ? The achievements are not very great in relation to the size of the task. One must, of course, acknowledge the work of the Commission and the E.E.C. Directives, which are sometimes a very mild version of the original proposals.

In 1978 the nine Member States consumed about 4.3 million tonnes of lubricants, which in due course led to some 2.1 million tonnes of "used oil". Half this amount was collected for salvage and there was uncontrolled dumping of the rest. Used oil is a potential secondary raw material that can be reprocessed and used almost in its entirety for other purposes or for incineration with energy recovery.

Only two countries, the Federal Republic of Germany and Denmark, had a well-organized system of collection and disposal. Since the E.E.C. Directive on used oil came into operation, the situation should have improved. But in July 1979 reminders on implementing the relevant directive were sent to Belgium, Italy, Luxembourg and the Netherlands.

The most important developments and endeavours often came from local initiative rather than from a new trend in central government policy. For instance the first selective refuse collections in Belgium came from the initiative of local authorities.

Recycling of paper is important because the E.E.C. Member States are dependent to the extent of 50 per cent on imported paper and wood fibre and the balance of trade deficit for this material is some Bfr. 320 000 million. It is the second highest amount, coming next after the deficit for imports of petroleum products.

Currently 9 or 10 million tonnes of wastepaper are being used in the E.E.C. to make recycled paper and board, which is about a third of Community consumption of paper products. This is a substantial achievement, representing 40 per cent use of wastepaper, but it is not enough and the Commission is proposing that the rate of use should be raised to 60 per cent.

Up to now the Community has approached the matter of refuse mainly from the point of view of pollution and its removal, but it now seems to be moving over, theoretically at least, to the idea that measures in that respect tackle only part of the problem of refuse and that more time should be devoted to preventing refuse.

In this connection, one wonders whether it is better to think about the best type of container for water from natural springs or to consider the more fundamental question of the actual principle of selling bottled spring water at all.

A Community map of recycling would show many worthwhile local initiatives and gaps for areas where nothing is being done. It would be a map of local efforts in managing refuse without any overall major policy.

The objectives to be attained might be identified as follows :

. reduction of the weight of containers ;
. recycling waste material and arranging selective collection ;
. reverting to some extent to returnable containers, and standardizing
  bottles ;
. using packaging materials which cause less pollution wherever
  possible ;
. not concentrating on packaging alone but seeking consumer goods
  that last or can be repaired.

An initial move in the right direction has been reduction of the weight
of containers : in 1950 a beer can weighed 83 g, now it weighs only 38 g

Recycling of waste material depends largely on arousing awareness in
consumers and in local authorities, but also private initiative where
the public authorities fail to act.

Reverting to returnable containers is desirable in many instances.
Someone recently said to me : "Surely you don't suggest having the
bottles from Perrier drunk in San Francisco brought back to France ?".
In that instance it would be absurd to insist on return, but not as
absurd as the outward journey of those bottles.

Return of containers would be less of a problem if manufacturers would
just be willing to standardize their containers to some extent.

Distributors do not like returnable containers and I do not think the
best way of dealing with this is to require the use of two kinds of
container, so that a non-returnable container allowing pollution and
a returnable container are used for the same product. The two kinds
will cause an undue increase in overheads and there is a great risk of
consumers choosing what they are persuaded to choose.

Producers or distributors will still impose their choice through
manipulation of prices, promotions and position on shelves. It would
be better to require the use of returnable containers for those products
with the shortest empty return journey.

How can we arrive at materials which cause less pollution ? It would
perhaps be difficult to prohibit particular packaging materials, even
if this may be the best method in some instances. The introduction of

a levy to pay for collection, treatment, pollution and the greater volume of imported raw materials might be a good idea, if the levy were sufficiently discouraging.

But there is a major difficulty here, because a tax imposed on the final consumer would not affect those whose decisions result in pollution. Also poorer people would be worse affected by tax than the well-off.

Lastly, a tax would cause complications to international trade. For example, if the Luxembourg authorities choose to finance part of the cost of refuse disposal from the proceeds of a tax, the tax could not be levied on production in France, Belgium or the Federal Republic of Germany. Goods would therefore have to be taxed on entering a Member State and the tax would have to be taken off goods leaving the territory.

It should be noted that as regards excise duty on petroleum products, neither the structures nor the rates have been harmonised. The Commission submitted to the Council a proposal on harmonisation of excise duty on mineral oils. This proposal, which has now reached the draft stage, actually confines excise duty on mineral oils to fuel oil and domestic heating oil. It is wrong that heating products, which are essential, bear an additional levy but petroleum by-products for packaging materials and outer wrappings are exempt.

Packaging materials may be a major source of pollution but they are only one instance of pollution caused by household refuse. For a long time consumers have been asking for longer-lasting and more repairable consumer goods. Two of the various ways of promoting these character-istics would be to reduce VAT on repairs and on secondhand goods to the lowest rate (relevant to the Member State which apply VAT on secondhand cars) or to adopt a directive on guarantees and after-sales services.

Finally, what is needed is effective measures from the Commission that will be bold and painful to the originators of pollution. This is necessary, because trade and industry are not capable of drawing up suitable measures on their own accord.

TOXIC AND DANGEROUS WASTE IN THE ECC

Dr. Benno W. K. Risch

Environment and Consumer Protection Service

Commission of the European Communities

SYNOPSIS

Toxic and dangerous waste arising out of industrial activities is a major environmental problem, and its treatment and disposal a number one qualitative problem of waste management.

20 million tonnes of such waste arise each year in the European Community, i.e. 15 to 20 per cent of industrial waste. Only in the medium and long term can we expect a gradual decline in the amounts as industry adapts itself to legal administrative and technical requirements, and as the extensive R & D produce results.

Specific directive exists on toxic and dangerous waste, to be implemented by the individual Member States by 20 March 1980.

Information available indicates that major provisions of the directive are already in force in at least six Member States. But as the directive is only a framework directive it will have to be supplemented and amplified.

For Phase 2 of the Community's endeavours in the toxic and dangerous waste sector the Commission has drawn up medium-term working programmes composed of two parts : "priority tasks and measures" and "other problems".

The Commission believes that the directive should be supplemented by priority by Community rules and provisions in the following areas :

- the drafting of a framework Lirective the temporary storage and dumping of toxic and dangerous waste, establishing uniform criteria and rules in respect of the selection, control and management of sites and dumps for such waste;

- the drafting of regulations on uniform labelling of toxic and dangerous waste and of containers for such waste, as well as the introduction of uniform danger symbols;
- the drafting of an implementing Directive on the documentary evidence of toxic and dangerous waste, dealing in particular with waste records, identification and accompanying documents for the carriage of such waste, particularly transfrontier carriage;
- special regulations on safety precautions in respect of hazards, accidents, etc.
- the setting up of a data bank on toxic and dangerous waste.

RESUME

Les déchets toxiques et dangereux issus des activités industrielles constituent un problème important sur le plan de l'environnement et leur traitement et leur élimination posent un problème qualitatif majeur en matière de gestion des déchets.

Il est produit chaque année dans la Communauté européenne 20 millions de tonnes de déchets de ce type, soit 15 à 20% des déchets industriels. Cette quantité ne diminuera progressivement qu'à moyen et à long termes à mesure que l'industrie se conformera aux prescriptions administratives et techniques légales et que les nombreuses activités de R & D porteront leurs fruits.

Il existe une directive relative aux déchets toxiques et dangereux qui doit être mise en oeuvre par les Etats membres pour le 20 mars 1980.

Selon les informations disponibles, des dispositions importantes de la directive sont déjà en vigueur dans au moins six Etats membres. Cependant, comme il s'agit d'une directive cadre, elle devra être complétée et amplifiée.

En ce qui concerne la deuxième phase des activités de la Communauté dans le secteur des déchets toxiques et dangereux, la Commission a établi des programmes de travail à moyen terme, subdivisés en deux parties : "tâches et mesures prioritaires" et "autres problèmes".

La Commission estime que la directive devrait être complétée en priorité par une réglementation et des dispositions communautaires dans les domaines suivants :

- la préparation d'une directive-cadre sur le stockage et la décharge temporaire des déchets toxiques et dangereux, fixant des critères et des règles uniformes de choix, de contrôle et de gestion des sites et des décharges destinés à ce type de déchets ;

- la préparation d'une réglementation relative à l'étiquetage uniforme des déchets toxiques et dangereux et de leurs récipients, ainsi qu'à l'introduction de symboles uniformes de danger ;

- la préparation d'une directive de mise en oeuvre concernant les références des déchets toxiques et dangereux, traitant notamment de l'enregistrement des déchets, des documents d'identification et d'accompagnement des déchets transportés, en particulier dans le cas du transport international ;

- une réglementation spéciale relative aux mesures de sécurité en cas de danger, d'accident, etc. ;

- la création d'une banque de données sur les déchets toxiques et dangereux.

Zusammenfassung

Giftige und gefährliche Abfälle, die insbesondere in der Industrie anfallen, stellen eines der wichtigsten Umweltprobleme dar, und ihre Verwertung und Beseitigung ist das vordringlichste qualitative Problem der Abfallwirtschaft.

Die Europäische Gemeinschaft produziert jährlich 20 Millionen Tonnen solcher Abfälle, was 15 bis 20% des gesamten Industrieabfalls ausmacht. Nur mittel- und langfristig ist mit einer allmählichen Verringerung dieser Menge zu rechnen, wenn sich die Industrie den Rechts- und Verwaltungsvorschriften sowie den technischen Regeln anpasst und die umfangreichen Forschungs- und Entwicklungsarbeiten Ergebnisse zeitigen.

Eine besondere Richtlinie über giftige und gefährliche Abfälle wurde verabschiedet, die von den einzelnen Mitgliedstaaten seit dem 20. März 1980 angewendet werden muss.

Aus den vorliegenden Auskünften geht hervor, dass die wichtigsten
Bestimmungen der Richtlinie zumindest in sechs Mitgliedstaaten bereits
in Kraft sind. Da es sich hierbei jedoch lediglich um eine Rahmenrichtlinie handelt, muss diese Richtlinie noch ergänzt und erweitert
werden.

Für die zweite Phase der Gemeinschaftsarbeiten im Bereich giftiger
und gefährlicher Abfälle hat die Kommission ein mittelfristiges Arbeitsprogramm erstellt, das aus zwei Teilen besteht: "Vorrangige Aufgaben
und Massnahmen" und "sonstige Probleme".

Nach Auffassung der Kommission sollte die Richtlinie durch Gemein-
schaftsregeln und -bestimmungen in folgenden Bereichen ergänzt werden:

- die Ausarbeitung einer Rahmenrichtlinie über die Lagerung und End-
  lagerung giftiger und gefährlicher Abfälle, einschliesslich einheitlicher Kriterien und Regeln hinsichtlich der Auswahl, Kontrolle und
  Pflege der Deponien für derartige Abfälle;

- die Ausarbeitung von Regeln über die einheitliche Bezeichnung von
  giftigen und gefährlichen Abfällen und ihrer Behälter sowie Einführung einheitlicher Gefahrensymbole;

- die Ausarbeitung einer Durchführungsrichtlinie über den Urkunden-
  nachweis im Zusammenhang mit giftigen und gefährlichen Abfällen,
  in der insbesondere das Verzeichnis, die Identifizierung und die
  Begleitdokumente für den Transport solcher Abfälle und vor allem
  die grenzüberschreitende Beförderung geregelt werden;

- besondere Regeln für Sicherheitsvorkehrungen bei Unfallgefahr und
  dergleichen;

- die Errichtung einer Datenbank für giftige und gefährliche Abfälle.

There is no doubt that the toxic and dangerous waste arising out of
industrial activities is one of the major environmental problems and
takes its place among the priority tasks of environmental policy.

Treatment and disposal of this toxic and dangerous waste is the number
one qualitative problem of waste management, not because of the quantities
involved but because of the particular hazards attached to it.

Some 20 million tonnes of such toxic and dangerous waste arise every
year in the European Community. It accounts for some 15 to 20 per cent
of the Community's annual arisings of industrial waste - an estimated
120 million tonnes at present. Most toxic and dangerous waste occurs
in the chemical industry as unavoidable by-products of industrial
processes.

In the last few years there has been an above-average increase in toxic
and dangerous waste as regards both quantity and their complexity. This
is, in part, a direct consequence of a heightened concern with the
environment, and is also due to advances in chemical and physical
knowledge and waste's threat to the environment. This has led to the
"discovery" of more and more new forms of dangerous waste.

A further disproportionate increase in the amount of dangerous waste
to be disposed of can be expected in the next few years, since
considerable additional amounts will arise through flue gas purification
at waste incineration plants, through sewage treatment and through the
fact that the dumping of waste at sea will be rendered more difficult
by international conventions.

Only in the medium and long term can we expect a gradual decline in the
amount of toxic and dangerous waste as industry adapts itself to legal,
administrative and technical requirements, as the cost of treating and
disposing of such waste rises and as the extensive research and development endeavours in this field begin to produce results.

Industry has already begun to react to these challenges by :

    (1)   replacing materials, processes and technologies, so as
             to reduce the amount of dangerous waste at the potential
             point of arising ;

    (ii)   introducing or applying low-waste technologies , this has
             largely eliminated hazards due to hardening salts ;

(iii)   increasingly recycling dangerous waste for energy purposes
or as secondary raw materials - for instance, following
galvanizing.

However, the steady increase in the quantity of toxic and dangerous
waste means that an optimum disposal technology is needed, which - on
land in particular - is able to suitably deal with such waste.
Particular efforts will have to be concentrated on improving technical
and organizational disposal and treatment methods.

For dangerous, as for other waste, there will have to be a switch time
from disposal to recylcing. There are three methods of waste disposal
and treatment : chemico-physical treatment, thermal treatment and
dumping. Optimized collection and transport systems, control and
monitoring regulations and the improvement of methods of chemical
analysis are important supplements to these methods.

Bearing in mind the growing extent of transfrontier haulage of such
waste between the Member States, the Action Programme of the European
Community on the Environment has assigned a high priority to this form
of waste.

In implementation of the programme and the framework Directive on waste
(15 July 1975) a separate Directive on toxic and dangerous waste was
adopted by the Council on 20 March 1978. This Directive lays down that
its provisions must be incorporated into national law by 20 March 1980
by the individual Member States.

Article 21 of the Directive requires the Member States to provide the
Commission by that date with detailed information on how the individual
provisions of the Directives were implemented. However, the Commission
does not yet have a conclusive review, but the information available
to us indicates that in at least six of the nine Member States major
provisions of the Directives are already in force.

The Directive on toxic and dangerous waste is of particular importance,
since it lays down common rules and provisions in respect of the major
problems of the production and disposal of toxic and dangerous waste in
this important and priority area of waste management.

The Directive contains the first definition at Community level of toxic
and dangerous waste in the form of an Annex listing the 27 most

important groups of toxic and dangerous substance to which the Directive is applicable.

The definition provided by the Directive – which was the subject of a very long tussle with government and industrial experts from the Member States – is by no means scientifically incontestable. To arrive at a scientifically incontestable definition, would have taken another ten years or more. The Commission therefore attempted to find a pragmatic solution in the form of a listing of substances.

The chief provisions of the Directives are :

(i)   the prohibition of uncontrolled discharge, uncontrolled transport and uncontrolled treatment and dumping of toxic and dangerous waste ;

(ii)   the establishment of appropriate labelling indicating the type, composition and quantity of waste ;

(iii)   the identification of sites at which toxic and dangerous waste is or has been – dumped and identification of such waste ;

(iv)   the requirement of a licence for plants, installations and undertakings which store, treat and/or dump toxic and dangerous waste ;

(v)   the requirement that the owners of toxic and dangerous waste who are not authorised to treat or dump such waste must hand over the waste to authorised plants, installations or undertakings for harmless disposal ;

(vi)   the requirement that the relevant authorities must draft and develop plans for the disposal of toxic and dangerous waste; these plans must provide for the necessary special treatment plants and suitable dumping sites; they must also be published ;

(vii)   the requirement that all plants, installations or undertakings which produce, own and/or dispose of toxic and dangerous waste must keep a special record of the quantity, type, physical and chemical characteristics, origin, method of disposal, dumping site and arrival and departure dates of such waste ;

(viii) the requirement that where toxic and dangerous waste is
transported in the course of disposal it must be
accompanied by a special identification form until its
final harmless disposal;   these forms must be preserved ;

(ix)   the requirement that every three years the Member States must
draw up a report on the disposal of toxic and dangerous waste
in their respective countries and forward it to the
Commission.

The Commission circulates this report among the other Member States and
reports to the Council and the European Parliament every three years on
the application of the Directive.   The first report has to be submitted
in 1981.   We hope that the Member States will inform the Commission
comprehensively and in good time, so that the Commission will be able
to comply with its obligations to the Council and Parliament.

The Directive of 20 March 1978 on toxic and dangerous waste in the
Community was initially designed as a specific implementing Directive
for the framework Directive of 15 July 1975.   However, in the course of
consultations with government experts from the Member States within
Council committees it became, in its turn, only a kind of framework
Directive for the toxic and dangerous waste sector.   A series of highly
specific rules in the original Commission proposal fell by the wayside
in the course of the consultations with the government experts.   This −
along with the complex nature of this waste − means that, as Phase 2
the framework Directive on toxic and dangerous waste will have to be
supplemented and amplified.

For Phase 2 of the Community's endeavours in the toxic and dangerous
waste sector the Commission, along with a working party of government
and industrial experts, has drawn up a medium-term programme.

This programme is composed of two parts :   "priority tasks and measures"
and "other problems".

The following points are listed among the priority taks and measures :

1.   Examination of identification and accompanying forms
relating to the transport of toxic and dangerous waste
pursuant to Article 14 (2) of the Directive, particularly
in respect of transfrontier carriage.

2.   Examination of documents relating to special regulations, particularly in respect of labelling, and to safety precautions in respect of hazards, accidents etc.

3.   Examination of criteria for the monitoring and management of the final dumping or longer-term dumping of toxic and dangerous waste. The drawing up of special storage plans.

4.   Transport problems.

5.   Examination of the question whether - and if so which - concentrations should be determined for the substances listed in the Annex to the Directive with a view to possible proposals to the Council.

6.   Particular problems of recylcing, treating and disposing of certain forms of waste, such as :

    (i)   organic halogen compounds ;

    (ii)   used solvents ;

    (iii)   waste derived from surface treatment ;

    (iv)   arsenic,

    for which special codes of practice are to be drafted.

Research and development endeavours, particularly the drawing up of a register of current and completed research and development projects run by the Member States and the Community, and identification of the research and development efforts are still needed in this high-risk sector.

The second part of the medium-term waste programme "other problems" covers the following points :

1.   Possible additions to the list of substances in the Annex to the Directive.

2.   An exchange of information and experience about central treatment plants.

3.   Packaging and conditioning (selection, criteria and regulations for containers, decontamination, disposal of used containers, etc).

4.   Insurance and liability.

5.   Recycling and re-utilization as secondary raw materials.

6. Replacement of materials and processes, particularly to reduce the production of toxic and dangerous waste.

7. Incineration and generation of energy.

8. Chemical, physical and biological treatment methods.

9. Waste which does not fall until now within the scope of the Directive.

As regards such waste not covered by the Directive, e.g. hospital waste, certain types of mining waste, etc., the idea is being considered of either drafting separate Directives or of harmonising the principles of existing or planned laws and administrative provisions of the Member States in this field by means of framework Directives, recommendations, guidelines and/or codes of practice.

The need for uniform identification and accompanying documents, in particular for transfrontier transport of this waste, arises from the fact that to an increasing extent toxic and dangerous waste is being carried across the national frontiers of the Member States. At present, control over toxic and dangerous waste ends at the national frontiers even of those countries which already have an elaborate control system.

From the point of view of the country of dispatch this waste disappears into outer darkness once it has crossed its frontier. However, control must be maintained up to final disposal.

Following many incidents which showed that transfrontier carriage is a particular source of danger the Commission takes the view that standard documents in several languages are required. This would at the same time facilitate border crossing which - because of highly divergent national regulations - is at present subject to highly cumbersome procedures.

Another important problem area, in respect of which the Directive of 20 March 1978 needs to be supplemented, is the transport of this waste. Since these substances are to an increasing extent being carried over great distances, there are particularly high risks involved, risks which are in many cases greater than these involved in the temporary and final dumping or treatment of this waste.

Although there are international conventions on the carriage of dangerous goods, they have still not been ratified and put into effect by all the

Member States. Furthermore, they apply only to international and not
to domestic transport. Some Member States have applied the rules of
these international conventions to domestic transport as well. However,
not all Member States have done so.

In fact some have adopted quite different regulations of their own,
which differ from the international conventions on international
transport.

Other Member States again have no specific regulations at all relating
to the carriage of dangerous goods. In other words, this matter is not
regulated uniformly within the Community. In view of the particular
risks in this sector and the many major accidents which have already
occurred, a uniform system is urgently needed within the Community.

The Commission is not necessarily aiming at a separate Directive on the
transport of toxic and dangerous waste. The matter could, for example,
be dealt with by extending existing regulations on the carriage of
dangerous goods to those substances and products which they so far fail
to cover. The main thing is that there should be a uniform Community
system covering the carriage of dangerous goods and that it should
satisfactorily settle the specific problems of toxic and dangerous waste.

Another important problem with which the Community has to concern itself
is the matter of concentrations of toxic and dangerous waste. The
Directive of 20 March 1978 contains no concentration values.

At present all the substances listed in the Annex to the Directive are
subject to the provisions of that Directive as regards, for example,
authorisation procedure, accompanying papers, control, etc., regardless
of the concentration in which they occur. Article 1 of the Directive
defines toxic and dangerous waste as follows :

" 'toxic and dangerous waste' means any waste containing or contaminated
by the substances or materials listed in the Annex to this Directive of
such a nature, in such quantities or in such concentration as to
constitute a risk to health or the environment".

Within the Community two Member States - Belgium and the Netherlands -
have set concentration values at national level. Two other Member States-
France and Britain - are at present drafting regulations setting out
concentration values.

Evidently other Member States have no intentions of setting concentration values for substances or materials contained in toxic and dangerous waste.

The Commission must see to it that the divergences between the Member States as regards levels of and criteria for concentration values do not impair full application of the Directive of 20 March 1978 or the proper working of the common market.

The Commission has therefore set up a working party of scientific and technical experts on toxic and dangerous waste, with the following tasks :

(i) to examine the basic problems, particularly in respect of the suitability of concentration values, basic criteria, scope of application etc ;

(ii) to collate and assess existing technical and scientific knowledge and available experience ;

(iii) to examine which concentration values might be proposed at Community level ;

(iv) to examine what effects - if any - on the working of the common market and the competitiveness of undertakings result from divergent national concentration values.

This working party will submit a report with appropriate proposals to the Commission in approximately a year.

The proper management of toxic and dangerous waste requires suitable treatment plants and special dumping sites. Article 12 of the Directive therefore prescribes that plans be drawn up covering the necessary special treatment plants and suitable disposal sites. This stipulation logically supplements the requirement that toxic and dangerous waste may be treated or disposed of only in authorised plants.

Because of the special dangers to man and the environment and in order to prevent water pollution and provide protection against nuisances, particular demands have to be made on sites and dumps where toxic and dangerous waste is temporarily or finally deposited. The selection, supervision and long-term safeguarding of sites and dumps for toxic and dangerous waste are therefore among the priorities of environmental policy and waste management.

A number of Community countries, the Netherlands in particular, have no –
or insufficient – sites with the required features. This means that
dumps and treatment plants must be accessible for waste from other
Community countries as well.

Toxic and dangerous waste often occurs in small and widely dispersed
quantities, which means that, for reasons of economic efficiency,
central facilities with an appropriate number of collecting points and
temporary dumps are needed. In the interest of economical, safe and
environmentally sound treatment and disposal of such waste, the areas
served by such central facilities within the Community have to be
supraregional and international.

There are a number of reasons why the Directive of 20 March 1978 must
be supplemented. In particular, priority Community rules and provisions
are needed in the following areas :

- the drafting of a framework Directive on the temporary
  storage and dumping of toxic and dangerous waste,
  establishing uniform criteria and rules in respect of
  the selection, control and administration of sites and
  dumps for such waste :
- the drafting of regulations on uniform labelling of
  toxic and dangerous waste and of containers for such
  waste, as well as the introduction of uniform danger
  symbols ;
- the drafting of an implementing Directive on the
  documentary evidence of toxic and dangerous waste,
  dealing in particular with waste records, identification and accompanying documents for the carriage
  of such waste, particularly transfrontier carriage ;
- setting up of a data bank on toxic and dangerous
  waste ;
- the drafting of uniform provisions in respect of
  safety rules covering hazards and accidents.

Toxic and dangerous waste derives mainly from chemical manufacturing
processes. In other words, they are closely related to industrial
activities. An economic consideration is that a whole series of toxic
and dangerous wastes which are at present destroyed are potential

highly valuable raw materials. Examples are the halogenated hydro-carbons, used solvents and the sludge from galvanizing processes.

All these considerations make research and development activities particularly important, and the Community should therefore set up a specific research programme on toxic and dangerous waste as part of its research policy.

In almost all the Member States and in the major industrialized countries long-term research is under way. There should however be far greater circulation of information regarding the subjects and results of research and testing activities. The setting up of a special data bank on toxic and dangerous waste therefore seems both useful and necessary. There should, in addition, be a regular exchange of experience and views on major topical or basic questions.

A brief exchange of thoughts and opinions on priority research topics and areas is due to be held during this very important conference on toxic and dangerous waste. The Commission would be grateful for suggestions; they could be used directly, as the Commission is at present drafting a third environmental research programme which is to take particular account of toxic and dangerous waste.

Frank, trusting and close cooperation between industry and the public authorities is of decisive importance to the success of a management policy for toxic and dangerous waste and this for several reasons :

In the first place, such waste derives mainly from industry. In the second place, treatment and disposal of such waste requires a highly developed technology.

Thirdly, the Commission takes the view in the light of the "polluter pays" principle accepted by the Community and of the liberal principles on which the Community is based, it is primarily industry's task to solve the problem related to the production, treatment and disposal of toxic and dangerous waste within the framework of conditions set by the authorities.

The task of the authorities is to exercise stricter control and super-vision in this particularly hazardous field and to establish the con-ditions required for safety by means of legal and administrative measures.

SOME VIEWS ON INDUSTRIAL WASTE

Yvan Cheret

National Federation of Waste Activities

France

SYNOPSIS

*The author looks at the situation from the point of view of private under-
takings specialized in waste collection and treatment, as opposed to
undertakings whose production process generates waste.*

*Great progress has been made in Europe in recent years and there are
centres where large amounts of waste can be appropriately treated.*

*Experience in running these centres has brought to light a number of
difficulties :*
  a) great variation in types of waste;
  *b) insufficient knowledge of their effects on man and the
     environment; this can give rise to conflicting reactions;*
  *c) wide variations in prices of treatment depending on the means of
     disposal selected;*
  *d) unfavourable development of the general economic situation as
     a result of the energy crisis.*

*This general situation explains why there are many and varied tensions
between the various groups involved in encironment policy: waste producers,
public opinion, administrations and political authorities. The waste
handling business itself is able to provide a range of solutions and can,
as has been proved, develop increasingly effective, but unfortunatley
often more expensive, processes. In fact, a large amount has already been
invested in the waste industry.*

*Since it is ultimately only the political authorities that can decide
which type of waste should be given which type of treatment, the waste
handling business requires these decisions to be clear, precise and*

long-lasting, in order that it can carry out its work correctly, directing its staff and investments towards solutions it considers to be right.

RESUME

L'auteur se place du point de vue des sociétés privées spécialisées dans la collecte et le traitement des déchets par contraste avec les sociétés dont le processus de production génère les déchets.

De grands progrès ont été accomplis en Europe ces dernières années et des centres existent qui permettent de traiter convenablement des nombreux déchets.

L'expérience de leur fonctionnement a permis de mettre en valeur un certain nombre de difficultés :

- *extrême diversité des déchets*
- *mauvaise connaissance de leurs effets sur l'homme et l'environnement ce qui peut provoquer des réactions contradictoires*
- *grande dispersion des prix de traitement selon la voie d'élimination choisie*
- *évolution défavorable de la situation économique générale sous l'effet de la crise de l'énergie.*

Cette situation générale explique l'existence de multiples tensions entre les divers acteurs de la politique de l'environnement : producteurs de déchets, opinion publique, administration, autorités politiques.

La profession du déchet, pour sa part, sait offrir une gamme de solutions et peut, elle l'a prouvé, développer des procédés de plus en plus efficaces, mais également, hélas, souvent plus coûteux. L'industrie du déchet est en effet une industrie déjà lourde en investissements.

Comme la décision sur le point de savoir si tel ou tel déchet doit faire l'objet de tel ou tel traitement, ne peut en finale être prise que par

l'autorité politique, la profession du déchet demande des décisions clai-
res, précises et durables pour lui permettre de faire convenablement son
métier en orientant ses hommes et ses investissements vers les solutions
considérées comme convenables.

ZUSAMMENFASSUNG

Der Autor vertritt die auf das Sammeln und Verwerten von Müll spezial-
isierten, privaten Firmen gegenüber den Firmen, in deren Produktions-
verfahren Müll anfällt.

In Europa sind in den letzten Jahren grosse Fortschritte gemacht worden:
es gibt Mullverarbeitungsanlagen, in denen zahlreiche Arten von Abfällen
vernünftig verarbeitet werden können.

Aufgrund der Erfahrungen mit diesen Anlagen können einige Schwierigkeiten
inzwischen richtig beurteilt werden:
- die extreme Unterschiedlichkeit des Mülls
- die unzureichende Kenntnis seiner Wirkung auf den Menschen und
  auf die Umwelt, was zu einander widersprechenden Reaktionen
  führen kann.
- der grosse Preisunterschied, der sich je nach gewählter
  Mullbeseitigungsmethode ergibt
- die ungünstige Entwicklung der allgemeinen wirtschaftlichen
  Lage unter dem Einfluss der Energiekrise

Mit dieser allgemeinen Lage lassen sich die vielfältigen Spannungen
zwischen den Interessengruppen innerhalb der Politik des Umweltschutzes
erklären,: müllproduzierende Unternehmen, öffentliche Meinung, Verwaltung
und Staat. Die müllverarbeitende Industrie inrerseits kann eine ganze
Reihe von Lösungsvorschlägen anbeiten und sie kann - das hat sie bereits
bewiesen - immer wirsamere, aber leider auch immer teurere Verfahren
entwickeln. In der müllverarbeitenden Inductrie sind hohe Investitionen
erforderlich.

Da die Entscheidung, ob diese oder jene Art von Abfällen auf diese oder jene Weise behenadelt werden soll, letzlich nur von den Behörden getroffen werden kann, fordert die Vertretung der Müllverarbeitenden Industrie klare, präzise und langfristige Entscheidungen; denn nor so kann sie ihre Arbeit ordentlich leisten und ihre Mitarbeiter und Investitionen in vernünftiger Weise lenken.

The waste industry which originated in small contractors with carts to remove sewage and household refuse has become very diversified and includes large transport organisations as well as operators of dumps and firms which build and manage treatment plants.

As far as industrial wastes are concerned, there are four distinct types of organisation which own and operate disposal plants. These are :

(i)   industrial manufacturers operating their own plants for treating waste they produce themselves ;

(ii)   a group of industrialists in the same area cooperating with a central treatment plant :

(iii)   government run treatment plants ;

(iv)   waste handling specialists taking the risk both of transporting others' wastes and of equipping their own specialized disposal plants, such as incineration or chemical treatment centres.

This text deals with the fourth category of waste handling specialists, which has developed quickly from 1970 to 1980 to a respectable size and become experienced and effective.

The plant of an individual unit may be designed for thermal treatment of mainly organic wastes, or for chemical treatment, that is to neutralize acids and bases and to fix heavy metals.

The capacity of these centres varies, but as a rule a reasonable sized unit - for either incineration or chemical treatment - deals with from 20 000 to 40 000 tonnes a year.

An investment of between 20 and 50 billion French francs is involved - depending on the size; this of course includes all general services, measurement and inspection laboratories etc.

Equipment existing in the NATO Member States in 1978 is mentioned in an article by Dr Bernd Wolbeck of the Ministry of the Interior of the Federal Republic of Germany.

In France, for example, about 20 specialized plants for the reception and disposal of industrial waste - run by companies as their sole concern - have been set up in the last six years, quite apart from the individual efforts made by manufacturers to treat their effluent, fumes and waste. About FFr 500 million have been invested in these centres and their local capacity is around 500 000 tonnes per year of toxic and dangerous waste.

The venture in France makes a significant contribution to the overall protection of the environment, and reflects the efforts of other countries.

However, companies running specialist industrial waste treatment centres are beginning to wonder about their future prospects.

Some lessons of experience are as follows :

1. Industrial waste occurs in very large quantities and in many different types. They can either be simple and unchanging, or highly complex mixtures which vary with time ;

2. Although the effects of a number of substances on man, animals and plants, are well-known, the effects of most industrial wastes – particularly medium- and long-term effects - are not known scientifically, which means that any hypothesis can be made and any position taken in speaking of their environmental impact.

3. It is more expensive than was previously thought, to break down various types of industrial waste into simple, harmless substances by thermal, physical or chemical means. One of the main reasons for this is the enormous variety of wastes received by treatment centres. Whereas the chemical industry knows what the proportions of different substances in the waste and keeps them at a fairly constant level, expressed in parts per million, operators of treatment centres cannot have an exact knowledge of the waste being treated, unless they carry out systematic tests which would double the price of treatment; nor is it possible to keep the composition of the waste to be treated constant for more than a few hours.

    Be that as it may, treatment of industrial waste involves a large amount of investment - about 1.5 to 2.5 times the annual turnover working at full capacity, as well as high maintenance costs.

4. Disposal costs vary a lot according to the types of waste and the means of disposal selected.

    Very roughly, the French figures are :
    FF 20 to FF 30 per tonne for normal disposal
    FF 70 to FF 100 per tonne for disposal with special facilities
    FF 250 to FF 700 per tonne for chemical or thermal treatment at on-shore plants.

In these circumstances it is normal for the company producing a
waste to ask the authorities for permission for it to be dumped,
not to mention the fact that some companies are tempted to get rid
of their waste illicitly.

5.   If disposal costs vary a great deal in absolute terms, their
      relationship to the value added or profits accruing from this
      type of production also vary a lot according to the processes used.
      In some cases these ratios are negligible, while in others they are
      appreciable. Taking an average rate for each of the main sectors
      is of no more than statistical significance, each individual factor
      having to reach its decisions in the light of its own circumstances.

      Whether or not the extra cost of suitable treatment can be passed
      on by the company to the consumer is another matter.

6.   The general economic situation does in fact govern policy decisions
      to a considerable extent.

      The concept of "environment" really caught on at the end of the
      1960s when the West was enjoying a period of economic wealth, and
      much effort has been deployed since then.

      Admittedly the energy crisis has faced governments with other
      priority tasks, which means that there is greater controversy than
      ever between public opinion - which is still highly aware of the
      need to combat pollution - and the authorities who are more
      concerned with dealing with unemployment and avoiding an energy
      shortage.

Waste handling specialists offer a range of effective ways of treating and
disposing of industrial waste. But they have to ask that the public
authorities should have clearly defined objectives in environment matters,
and to maintain them over a long period, in order to avoid investments
which remain unused.

All waste which is not recycled economically should clearly be reinte-
grated into natural environment : the problem is to do this without
damage and with a minimum of inconvenience. This means for each type of
waste there should be indications of a number of places where it can be
received, and of the form in which it can be received.

The public authorities' present idea seems to consist in drawing a distinction between three main types of waste :

a) "ordinary" wastes which can be dumped without any apparent difficulties, subject to precautions which could be termed "elementary" ;

b) "special" wastes which present a number of risks, but which can be dumped on condition that the site is genuinely suitable, and subject to strict permanent supervision ;

c) "toxic" or "dangerous" wastes which cannot be dumped and must undergo physio-chemical treatment, by means of incineration etc., to be converted into "ordinary" or "special" wastes.

The industry approves of this analysis which can make a basis for sensible policies.

Note that each potential dumping site has its own characteristics. The suitability of a site for a certain type 'of waste depends on the results of hydrogeological study. Such studies may show the need for work to be carried out (such as making the site watertight) before operations can begin.

A future difficulty lies in the definition of the wastes suitable for each of the three methods : the two types of dumping and treatment (chemical or thermal etc.).

This definition should be as easily understood by the producer of wastes as by those treating it and by the inspection bodies, bearing in mind that staff responsible for its everyday application are often relatively untrained.

The lists of products should therefore be very clear, all terms must be used with the most logical meaning and that corresponds to normal usage; explanations should be written clearly and comprehensibly. The same product should not appear in two different categories.

Wherever possible, simple tests for deciding what to do with waste and for checking this decision, should be described.

When it is not desirable to have a certain type of waste - ordinary or special - in a dump, there is the problem of deciding whether it should be banned completely or whether dumping standards should be drawn up, and, if so, whether this should be expressed in terms of the content or of permissible overall quantities.

Regulating in terms of content tends to suggest carrying out dilution. This method is hardly to be recommended for dumps with a limited capacity.

One might envisage fixing overall quantities of e.g. chromium - which might be permitted on a dump before having to close the dump.

The industry considers that certain products should be banned outright in view of the difficulties of implementation and inspection, the risks of overlapping responsibilities and of disputes. Prohibited products could be listed in ministerial orders which would state that exceptions could be made - on the merits of the individual case - only if there were no other solution (if treatment centres were saturated), on the written instructions of the supervising authorities for each case.

Supervision

Carrying out the supervision in a responsible manner is the keystone of the whole operation. It is only this that can guarantee the effectiveness of the efforts of industrialists that produce special wastes and set up treatment centres, and of those who agree to pay a high price for dis- posing of their wastes. Heavy penalties should be imposed for failure to comply.

Responsibilities should be clearly defined, such as :

a)   responsibility of the waste producer for defining the products to be taken away ;

b)   the responsibility of the operator, taking into account statements by the producer and the available means of checking ;

c)   responsibility of the administration when it indicates the destination of a given product.

Necessary tests and degrees of responsibility should be clearly defined since it is not possible to test every load arriving at a dump.

As is generally the case at present, where government policy leads to the setting up of a number of specialized plants for the treatment of industrial wastes, several questions arise with regard to making the requisite investments.

The allocation of public appropriations is a difficult matter. One of
the dynamic features of the Western economy seems to be the fact that
there are flexible financial mechanisms which allow decentralized
schemes to develop freely.

Basically, wherever there is a market for a given product or service,
there will be someone to supply this product or service.

This applies equally well to our business, where collection facilities,
dumps and transit stations have been created privately as soon as a real
need has appeared.

This dynamism is strikingly illustrated by the recent development of
waste treatment plants. A concesus of opinion on the fact that "toxic"
substances could not simply be left to run off just anywhere, was enough
to conjure up investors to summon up financial backing and to create
plants.

Does all this mean that public financial aids are not necessary in the
field of industrial wastes ? We do not think so, because great changes
have to be made, but we are convinced that these interventions will be
effective only if the points at which they are applied and the mechanisms
used are geared to the difficulty of the problems involved.

The waste industry is convinced that it is economically viable for waste
producers to bear the cost of burying the waste at ordinary dumps,
financed entirely by private means, but that it is legitimate to provide
public financial aid for more expensive treatments, in order to reduce
the temptation to defraud.

Our first suggestions is then to direct public aids towards the more
dangerous products and those which require the most expensive treatment.
It seems justifiable to provide aids for chemical or thermal treatment
plants, and even dumps for special wastes.

Apart from providing financial aid, should the public authorities, in
certain cases, also be responsible for setting up treatment plants ?
This seems desirable neither in theory nor in practice. Let each man
do his own job, with the public authorities laying down a clear policy
and the waste disposal companies seeing that their plants function
properly.

WASTE DISPOSAL BY THE CHEMICAL INDUSTRY

Dr James T. Farqhar

Albright & Wilson

Great Britain

SYNOPSIS

Chemical industries in EEC member states dispose of about 45 million tons of waste per year. This quantity is relatively small in terms of total waste disposals in these countries, but the difficulty of disposing of chemical wastes is complicated by their great diversity of composition and by the fact that some of them are toxic or otherwise dangerous.

Consideration of the various means which are available for the disposal of chemical wastes indicate a wide cost range. Where disposal to land is acceptable, its cost is usually small compared with the other costs of the industry. Unfortunately, it can be readily shown that the wide scale adoption of expensive waste treatments could have a dramatic effect on the economics of chemical manufacture and would have a serious impact on the prosperity of the industry and on the employment which it provides. This leads to the inescapable conclusion that practices for the disposal of waste which are to be enforced or encouraged by law, must be most carefully examined to ensure that they represent the best possible option in each particular case. Policies "to be on the safe side" whichever side of the argument is taken as "safe" must be considered to be unacceptable.

As the potential problems of disposing of chemical wastes have been realised over the last half century or so, a great deal of sophisticated research has been conducted on them. An example of this has been the build-up of expertise in recent years on the behaviour of different classes of wastes in landfill sites so that it is now highly improbable that any land-based disposal carried out under currently established good practice, will give rise to environmental problems at any time in the future. The ideas expressed several years ago that the disposal of chemical wastes to land

*would eventually be phased out have no place in current knowledgeable
thinking.*

*In examining these problems, one of the difficulties being faced is the
tendency for fallacious assumptions to creep in. It is quite wrong to
suppose that the disposal of waste is necessarily bad. There is a cost
associated with the safe disposal of any particular class of waste and
this cost completely defines the extent of the problem.*

RESUME

*Les industries chimiques des Etats membres de la CEE éliminent 45 millions
de tonnes environ de déchets par an. Cette qualité est relativement faible
par rapport au total des déchets éliminés dans ces pays, mais la difficulté
que soulève l'élimination des déchets chimiques est accrue par leur grande
diversité de composition et par le fait que certains de ces déchets sont
toxiques ou dangereux.*

*L'examen des divers moyens dont on dispose pour éliminer des déchets
chimiques fait apparaître un large éventail de coûts. Lorsque l'élimination dans le sol est acceptable, son coût est généralement faible par
rapport aux autres coûts de l'industrie. Malheureusement, il est aisé
de montrer que l'adoption à grande échelle des traitements coûteux des
déchets pourrait avoir un effet désastreux sur la rentabilité de l'industrie chimique et aurait un grave impact sur la prospérité de cette industrie et sur les emplois qu'elle procure. On aboutit donc à la conclusion
inéluctable que les pratiques en matière d'élimination des déchets qui
doivent être imposées ou encouragées par la loi doivent être examinées
avec un soin extrême pour s'assurer qu'elles constituent la meilleure
option possible dans chaque cas particulier. Il convient de considérer
comme inacceptable des politiques visant à se "mettre à couvert" quel que
soit l'aspect du problème qui est jugé être "à couvert".*

*On s'est rendu compte au cours du dernier demi-siècle des problèmes
potentiels que pose l'élimination des déchets chimiques, un grand nombre
de recherches très poussées ont donc été menées dans ce domaine. Un
exemple nous en est fourni par les connaissances acquises au cours des
dernières années sur le comportement de différentes catégories de déchets
dans des sites de remblai, de sorte qu'il est actuellement très improbable*

qu'une élimination dans le sol selon les pratiques actuellement bien éprouvées soulève des problèmes d'environnement à l'avenir. Les idées exprimées il y a plusieurs années selon lesquelles l'élimination de déchets chimiques dans le sol serait dépassée par la suite n'ont pas cours dans la théorie actuelle.

L'une des difficultés que l'on rencontre lors de l'examen de ces problèmes est la tendance des hypothèses fallacieuses à fausser le raisonnement. Il est tout à fait erroné de supposer que l'élimination des déchets est nécessairement mauvaise. L'élimination sûre d'une catégorie particulière de déchets implique un certain coût qui délimite nettement l'étendue du problème.

ZUSAMMENFASSUNG

Bei der chemischen Industrie in den E G-Mitgleidstaaten fallen pro jahr etwa 45 Millionen Tonnen Adfall an. Im Vergleich zur gesamten Abfall-beseitigung in diesen Ländern ist diese Menge Verhältnismässig gering, die Schwierigkeit
der Beseitigung von chemischen Abfällen wird jedoch dadurch erhöht, dass diese Abfälle ihrer Zusammensetzung nach sehr unterschiedlich sind und dass einige von ihnen toxisch oder auf andere Weise gefährlich sind.

Dir verschiedenen Methoden für die Beseitigung von chemischen Abfällen bewegen sich innerhalb einer breiten Kostenspanne. Sofern die Abfall-beseitigung in einer geordneten Deponie zulässig ist, sind die entsprechenden Kosten im Vergleich zu den übrigen Kosten der Industrie im allgemeinen gering. Leider ist es jedoch so, dass sich der weit-gehende Rückgriff auf kostspielige Abfallbehandlungen in dramatischer Weise auf die Wirtschaftlichkeit der chemischen Industrie auswirken und Rückschläge auf die Konjunktur der chemischen Industrie und auf die in diesem Industriezweig bereitgestellten Arbeitsplätze habben könnte. Dies führt zu der Schlussfolgerung, dass Praktiken der Abfallbeseitigung, die nicht durch das Gesetz vollstreckt oder gefordert werden, ausser-ordentlich sorgfältig geprüft werden müssen, damit gewährleistet wird, dass sie in jedem einzelnen Falle die optimale Lösung darstellen. Eine einfach Politik das "Auf-Nummer-Sicher-Gehens" - was immer dabei als sicher gelten wag - ist als unzulässig zu betrachten.

Im Laufe der letzten fünfzig Jahre ist man sich der potentiellen Probleme im Zusammenhang mit der Beseitigung chemischer Abfallstoffe bewusst geworden, und es wurden zahlreiche differenzierte Forschungs-

arbeiten über diese Fragen durchgeführt. Ein Beispiel hierfür war in den letzten Jahren das Erlangen von Fachwissen über das Verhalten verschiedener Arten von Abfall in Deponien; es ist daher heute höchst unwahrscheinlich, dass eine nach dem haute anerkannten ordnungsgemässen Verfahren durchgeführte Ablagerung in einer Deponie in der Zukunft Probleme des Umweltschutzes aufwirft. Die vor einigen Jahren zum Ausdruck gebrachte Vorstellung, dass die Ablagerung von Chemieabfällen in Deponien letzlich eingestellt werden müsse, ist nach dem heutigen fortgeschrittenen Stand der Kenntnisse nichtmehr haltbar.

Eine der Schwierigkeiten auf die man bei der Prüfung dieser Probleme stösst, ist die Gefahr, dass sich irrige Postulate einschleichen können. Man darf Reinesfalls davon ausgehen, dass die Abfallbeseitigung notwendigerweise schlecht ist. Die sich ere Beseitigung jeder Art von Sondermüll verusacht Kosten, und der ganze Umfang des Problems wird eben durch diesen Kostenfaktor umrissen.

Die Geschichte des Menschen auf der Erde war von jeher weitgehend davon bestimmt, mit welchem Erfolg es dem Menschen gelungen ist, die natürlich-en Rohstoffe unserer Erdoberfläche für seinen Unterhalt und sein Wohl-befinden zu nutzen.

Nach dem Gesetz der Erhaltung der Masse muss für jeglichen Stoff, den der Mensch aus einer ihm nutzbaren feststehenden Anhäufung von Material entfernt, genau die gleiche Menge an Material auf irgendeine Weisse und irgendwo wieder in die natürliche Umwelt eingebracht werden.

Much muddled thinking continues on the subject of waste production and waste disposal. The reason is that few countries had any formal legislation specifically controlling waste disposal until about 10 years ago.

At that time, a wave of enthusiasm for protecting the environment spread throughout the advanced countries and threw into relief the fact that waste disposal was largely uncontrolled. This, coupled with publicity for a very small number of cases where waste disposal had subsequently given rise to significant damage to the environment and health, has caused an over-reaction. Most advanced countries are now in the process of proposing and introducing bureaucratic controls and expensive schemes for administration far greater than can be justified by the environmental risk.

To assess the potential dangers of waste disposal operations by assuming all sorts of horrific happenings if cyanides or mercury compounds get into town's water supply may make good copy for newspaper editors, but as a sound basis for deciding what is necessary to protect the environment they are of little practical value. The fact remains, that mankind, particularly in the European area, has been producing increasing quantitiesof dangerous waste materials for upwards of 200 years and has been disposing of them in a completely hap-hazard and uncontrolled way until very recently. Surely with the advantage of this past experience, it must be possible to assess the real practical dangers associated with waste disposal and then take steps to see that these real practical dangers are obviated.

It can already be stated with a considerable degree of confidence, that where management of waste disposal is now being conducted within the confines of up-to-date established scientific practices, it is exceedingly improbable that environmental damage, even in the very long term, will be encountered. I will have more comment to make on the administration of waste disposal later. In the meantime, I would like to interpose at this point a matter which my scientific colleagues and I in industry feel is not sufficiently understood about the fundamental reasons for waste arisings.

Man's basic activities on the surface of the earth are to remove materials from the natural environment, to use them for his own convenience and well-being, and then return them to the natural environment whence they came. Nature rarely provides materials in a pure state and usually, on taking materials from nature, the first thing that man has to do is to throw away part of the material which he does not require. When a man pulls an orange from the tree the first thing he does is throw away the skin and when he digs a mineral out of the ground, he throws away the shale or other useless material which nature has mixed with it.

The material which man takes from nature usually has to be processed before use. At this stage, quite inevitably, more waste is produced. Much industrial waste leaves the system at this point. Now we pass to the use of materials. This is where man eats the food, wears the clothes, and drives in the motor cars and then eventually these materials wear out.

Inevitably again, waste is produced although at this stage, some material can be recycled back to the beginning of the processing stage and so its ultimate return as waste to the natural environment is delayed.

The disposal of waste is a fundamental necessity of modern living. There is nothing at all immoral about it and the great majority of waste is disposed of without any real significant or even potential damage to the environment. From this it follows that general concepts such as the development of no waste technologies or clean technologies represent a waste of effort based on fallacious thinking. Where real difficulties can be demonstrated to arise because of specific waste disposal operations, these must be looked at as individual cases and if, in a very few instances, it looks as if the best option may be to change manufacturing techniques, then so be it. The over-riding requirement must however, be to identify such problem wastes and not to spend money in areas where no problem arises.

Since legislation specifically controlling waste disposal started to come in about 10 years ago, progress towards a complete and permanent system covering different classes of waste has been much slower than was originally intended. With the possible exception of Germany, many proposals made years ago have yet to be implemented in full. This is not only true at national levels, it is also so at EEC level. Some of the provisions on waste disposal laid down in the 1973 edition of the "Programme of Environmental Action for the European Communities" have yet to be implemented in directives.

My interpretation of the delay in finalising control systems in most
countries is that it is in the final stages of working out the details
of such systems and in the fitting of the meat onto the bones so to
speak, that the greatest administrative expense and cost for industry
arises. I believe that all over the place, administrators are engaged in
examining administrative systems, finding that their introduction will
increase public expenditure and are seeking in vain for the environmental
advantage to justify this. The fault of course lies with the politicians
in the early 1970s who indulged in politics of expediency and then left
their successors to suffer the consequences and indeed to suffer the
embarassement.

I must say that I believe that the organisers of this seminar are to be
congratulated on their sense of timing because this is certainly a time
to sit back and take a fresh look at the relationship between waste
disposal and the environment and I would like to look in more detail if
you will allow me, at one or two of the mistakes which I believe have been
made.

Perhaps the most basic error in the drafting of waste management control
has been that attempts have been made to make it all-embracing and to
cover far too many operations. The obvious source of environmental danger
from waste disposal involves the improper use of landfill sites. It is
right and proper that the use of landfill sites for the permanent disposal
of waste should be subject to a licensing system and such is a feature of
the legislation in most countries. Recently, a great deal of work has
been in progress involving scientists in several disciplines in which the
behaviour of different wastes disposed of in different types of landfill
sites have been examined both in the field and in deliberate laboratory
experiments. The importance of this work cannot be too strongly
emphasised as its outcome will undoubtedly indicate under what circumstances various quantities of different classes of waste can be safely
landfilled on any particular site. The award or refusal of site licences
for landfill sites based solely on sound scientific data should form the
basis of all controls on waste disposal. It is important that once a
landfill site has been investigated to establish the load which can safely
be disposed of to it, its use should be enjoyed freely by those whom the
owners authorise to use it always provided that the licence conditions

are met.

It should be the duty of each controlling authority to monitor the sites
which it licences to ensure that they are being operated within the law.
Those who deliberately flout the law by disposing of wastes in an irres-
ponsible way should be brought to book and the courts should impose severe
penalties on them. It is unfortunate however, that the development of
legislation on waste disposal in recent years has, in addition, included
extensive paperwork under which those disposing of waste must expend
time in notifying the authorities on the details of their day to day
operations. It is exceedingly doubtful if anything is thereby gained.
It is highly unlikely that anyone who proposes to dispose of wastes outside any licence requirement will effectively write to the authorities
telling them of his actions.

Unfortunately, in their enthusiasm for controls, most governments and
indeed the EEC, also require lice n sing for waste treatment plants. This
is complete nonsense. There are no statistics to indicate that the
operation of a plant treating waste is any more dangerous to the
environment than any other manufacturing plant. The one thing that is
absolutely certain about any waste treatment plant is that for every ton
of waste that goes into it, at least one ton of material has got to come
out of it for disposal at the other end. I say at least one ton because
almost always the treatment process will involve the addition of other
materials required to bring about the treatment and this too has to be
disposed of at the end of the process.

When the inevitable effluent from a waste treatment plant comes to be
disposed of, control is of course, essential. This may be material
going to a landfill site in which case the provisions that I have already
discussed must apply. Often, the effluent will be in an aqueous form
going to a natural water course or it may be in a gaseous form going up
a chimney to the atmosphere. In each of these cases, controls are
properly exerted by the appropriate inspectorate and these are well
established in all EEC member countries. Lastly, other controls to ensure
the prevention of nuisance to those living in the area, apply just as
much to waste treatment plants as to any other kind of plant. In short,
the control of waste treatment plants under waste disposal legislation
is an irrelevance which should certainly receive the close attention of

any of our politicians who are at all serious in trying to reduce bureaucracy and public expenditure.

At this point, I would like to take a moment to point out that the sole purpose of a waste treatment plant is to convert a waste material which may seem difficult or dangerous to handle or to dispose of directly back to the environment, into a different waste material with more favourable properties. As I have already pointed out, this will usually involve an actual increase of the amount of waste to be disposed of. It will almost always involve a process which is itself wasteful of resources and will usually be wasteful in its use of our increasingly scarce and expensive supplies of fuel.

One aspect of waste disposal worth mentioning is the drawing up of codes of practice specifying recommended means by which different classes of waste should be disposed of or in other words, the means by which types of waste should be returned to the natural environment. The process of drawing up codes of practice is at a very advanced stage in the United Kingdom having been commenced about seven years ago due to the foresight of a new retired civil servant, Mr Arthur Kenny. We in the U.K. believe that these codes of practice are valuable and while they do not carry the force of law, they were all drawn up by committees sitting in Department of Environment offices and so carry the status of government recommendations. Some of these U.K. codes have been on issue for a number of years and there has been little criticism of them. Although not everything in the U.K. codes of practice may be acceptable to the European Commission in its current efforts to produce codes of practice for the EEC, it seems likely that they will form a useful basis for discussion.

The same kind comments cannot be made about the London and Oslo Conventions under the terms of which the depositing of wastes at sea by dumping from ships and aircraft are controlled. These international agreements were drawn up and signed about 8 years ago at the behest of a number of politicians who thought that by so doing, their electoral chances might be enhanced. Unfortunately, although most of these politicians have now departed elsewhere, the agreements remain and their loyal application by those whose job it is to do so probably does as much damage to the environment as it does good. These agreements carry a black list of classes of materials which may never be deposited in the sea under any circumstances. There is a general prohibition on all compounds of mercury and cadmium, a requirement which almost certainly increases environmental danger rather than decreases it. If it is necessary to dispose of a waste containing either mercury or cadmium and if, for some reason, it is impracticable to recover values from that waste, the safest environmental alternative is almost certainly dispersion in the waters of the deep ocean. There, the addition of even the world's total commercial usage of these metals and their compounds would add but imperceptibly to the quantities of these elements already present. Deep sea disposal is not cheap and its use for special wastes would never be widespread. No one would therefore advocate the widespread use of ocean dispersal as a preferred policy, but it does nothing to aid confidence in officialdom to find the best environmental option being subject to total prohibition.

QUESTIONS & COMMENTS

Mr Menke-Glückert made the points that :

> *(i) though he was generally in favour of a reduction in red tape ... "I have some doubt that the area of toxic waste is an area to try working it out."*

> *(ii) every encouragement should be given to industry to change from practices involving just throwing away material and replace them with recycling.*

Dr Farqhar emphasised that to reapcess materials where it was not economic was to walk the road to bankruptcy. He agreed with a questioner that landfill sites should be rigorously controlled.

Discussing the harmonisation of legislation, he thought that one could not have the same legislation applying to, say, 100 tonnes of salt being discharged into the Upper Rhine, which should be opposed, as to the same salt being put into the sea off the West Coast of Scotland, where it would have no effect.

THE USE OF WASTE IN AGRICULTURE

Léon Klein

Environment and Consumer Protection Service

Commission of the European Communities

SYNOPSIS

*The present situation of fermentable-waste management in respect of legis-
lation and organization in the Member States is described. The problems
posed by the use of compost and sludge in agriculture will be chiefly
discussed.*

*As regards compost, the quantities produced in the Member States are,
generally speaking, very small, if not almost nil. The largest producers,
and hence users, are the Federal Republic of Germany, the Netherlands and
France, where 2 to 40 % of domestic waste is composted. In no Member State
do we find legislation dealing specifically with compost. However, the
law is more thorough-going in the three countries mentioned, although it
differs considerably from one to another. And it deals only with the
handling of waste and the making of compost. The question of how compost
affects the soil does not appear to be reflected in the legislation of
the Member States.*

*As regards sludge, the quantities produced in the Member States amount at
present to 6 million tonnes of dry matter per year, i.e. 800 kgs per
person. But only 29 % of the sludge produced is used in agriculture, while
64 % is dumped either on land or at sea. The fact that there is reluctance
to use sludge as a fertiliser is mainly due to the problems of pollution
and poisoning it could cause, for fresh, that is to say untreated, sludge
is a source of pathogens and can transmit soil and plant diseases.
Furthermore, if industrial effluent is treated along with domestic effluent
the sludge may contain excessive quantities of heavy metals which are toxic
to crops and thus to living organisms.*

*In some countries there are already a number of legal provisions relating
to sludge used in agriculture - usually concerned with the maximum permissible content of heavy metals, although such limits are expressed in
different ways and sometimes vary considerably.*

-117-

*In view of this the Commission intends to put up a proposal for a
Community directive in the near future which will aim mainly at better
utilization of sludge while limiting its application in certain cases
when it has not been treated and has too high a heavy-metal content.
Certain conditions of treatment and of analysis of sludge and soil will
have to be satisfied before any use is made of sludge.*

RESUME

*L'utilisation   des déchets en agriculture permet, d'une part de résoudre
de manière substantielle le problème de l'elimination des déchets, et,
d'autre part, elle peut être d'un intérêt non négligeable pour l'agriculture. Si de grandes possibilités existent dans ce domaine, l'utilisation
des déchets en agriculture comporte néanmoins certains aspects négatifs
(pathogènes, métaux lourds ...) Il emporte donc de préserver, tant à
moyen qu'à long terme, notre environnement ainsi que les conditions de
production les meilleures pour l'agriculture. C'est pourquoi l'action du
groupe de travail développe ses efforts dans les directions suivantes :*

*1. Faire le point des expériences tendant à promouvoir l'utilisation
agricole des déchets dans la C.E.E. et les pays tiers;*

*2. Examiner les possibilités actuelles et potentielles d'une telle utili-
sation en tenant compte des aspects techniques, économiques, juridiques
et écologiques du problème;*

*3. Recommander des mesures communautaires adéquates notamment dans les
domaines suivants :*

    - normes visant la qualité des produits organiques
    - organisation de la commercialisation
    - recherche et développement.

*Une première réunion du groupe de travail "Utilisation des déchets en
Agriculture" du C.G.D. s'est tenue à Bruxelles le 10 juillet 1979. Outre
la présentation de l'etude "L'épandage des effluents délevage sur les sols
agricoles dans la"Communauté Européenne", l'accent de cette réunion a été
surtout porté sur les boues d'épuration et a finalement abouti aux conclusions suivantes :*

  - La caractérisation agronomique des boues utilisables est un facteur

essentiel pour l'utilisation en agriculture. Cette caractérisation ferait état tant des aspects positifs (azote, phosphore, matière organique) que des aspects négatifs (métaux lourds, oligoéléments, pathogènes ...).

- Afin de parler le même language, une liste des anayses indispensable à la caractérisation agronomique des boues, de même qu'une standardisation des méthodes d'analyses semble nécessaire au niveau de la Communauté, ceci purrait faire l'objet d'une norme européenne, reprise, le cas échéant, par la suite, sous forme de directive.

- L'utilisation des boues d'épuration en agriculture doit faire l'objet de deux types de précaution :
  . Le 1er concerne les boues proprement dites où des précautions sont à prendre dans les domaines tels que stockage, transport, traitement, odeurs, pathogènes, oligo-éléments, destinations ...
  . le 2ème concerne la protection des sols; il conviendrait, en effet, de limiter l'accumulation d'éléments indésirables dans les sols.

Toutes ces condidérations expliquent le fait que notre action s'est particulièrement concentrée sur l'utilisation des boues en agriculture comme nous le verrons dans la suite de l'exposé.

Cet exposé traitera successivement :
- du compost
- des effluents d'élevage
- des boues d'épuration.

La situation dans les Etats membres sera évoquée.

ZUSAMMENFASSUNG

Die Nutzung der Abfälle in der Landwirtschaft kann in erheblichem Masse sur Lösung des Problems der Abfallbeseitigung beitragen und kann bei guter Beratung und Anleitung durchaus von Interesse für die Landwirtschaft sein.

Die heutige Lage in der Bewirtschaftung gärfähiger Abfälle wird unter dem Aspekt der Rechtsvorschriften und der Durchführung in den Mitgliedstaaten der EG beschrieben. Vor allem geht es um Probleme bei der Verwendung von Kompost und Klärschlämmen in der Landwirtschaft.

In den Mitgliedstaaten der Gemeinschaft wird meist äusserst wenig,
manchmal sogar gar kein Kompost erzeugt. Noch die grössten Erzeuger und
entsprechend Verwender sind die Bundesrepublik Deutschland, die
Niederlande und Frankreich, wo 2 bis 40 % des Hausmülls durch
Kompostierung behandelt werden. Speziell auf Kompost bezogene
Rechtsvorschriften sind in keinem Land zu finden. Allerdings befassen
sich die Rechtsvorschriften in den drei genannten Ländern stärker mit
dieser Materie, auch wenn die Unterschiede von Land zu Land gross sind.
Überdies betrifft die rechtliche seite nur den Umgang mit den Abfällen
und die Kompostherstellung. In den Rechtsvorschriften der
Mitgliedsländer wird der Aspekt "Einfluss des Komposts auf den Boden"
offensichtlich nicht hervergehoben.

Die menge des in den Mitgliedstaaten erzeugten Klärschlamms erreicht
gegenwärtig 6 Millionen Tonnen Trockenmasse, das entspricht 800 kg je
Einwohner. Von den insgesamt anfallenden Klärschlämmen werden jedoch
lediglich 29 % in der Landwirtschaft verwendet, während 64 % auf dem
Festland bezw. im Meer abelagert werden. Der Grund fur den zögernden

Einsatz der Klärschlämme als Dünger liegt vor allem darin, dass Umweltverschmutzung und Vergiftungsprobleme auftreten können; frischer
Schlamm, der nicht behandelt wurde, birgt Krankheitserreger und kann
somit bestimmte Krankheiten für den Boden und die Pflanzen hervorrufen.

In einigen Ländern bestehen bereits Rechtsvorschriften über die in der
Landwirtschaft genutzten Klärschlämme. Sie beziehen sich meistens auf
den Grenzwert für den Gehalt an Schwermetallen, wenn auch der Ausdruck
dieser Grenzwerte unterschiedlich ist und die Grenzwerte oft stark von
einander abweichen.

Daher will die Kommission in Kürze einen Vorschlag für eine
Gemeinschaftsrichtlinie vorlegen, mit der vor allem die Klärschlämme
besser genutzt werden sollen, während gleichzeitig ihre Verwendung in
bestimmten Fällen begrenzt wird, wenn die Schlämme mit zu hohem
Schwermetallgehalt nicht behandelt wurden. Vor jeder Nutzung sind
bestimmte Behandlungen sowie Analysen der Klärschlämme und des Bodens
erforderlich.

A Working Party has been set up under the Waste Management Committee to examine the possibilities of using waste in agriculture. The waste in question may be grouped under the general heading of "fermentable organic material" - more particularly compost, effluent from stock rearing and sewage sludge.

The use of this type of waste in agriculture will go a long way towards solving the problem of waste disposal and is also of considerable value to agriculture. Although there is great scope for the use of waste in agriculture, it also has its drawbacks (pathogens, heavy metals, etc.). It is essential in both the medium and long term to protect our environment and provide the best possible farming conditions, and it is for these reasons that the Working Party has decided to :

1. carry out a survey of experiments in Community and other countries to promote the use of waste in agriculture ;

2. study existing and potential possibilities, having regard to the technical, economic legal and environmental aspects ;

3. recommend suitable Community measures on :
   a) quality standards for organic products
   b) marketing organisation
   c) research and development.

The Working Party held its first meeting in Brussels on 10 July 1979 at which it discussed the study on the spreading of animal excrement on utilised agricultural areas of the Community, and sewage sludge in particular, and came to the following conclusions :

(a) The classification for agricultural purposes of suitable sludges is essential. Both their good (nitrogen, phosphorus, organic matter) and bad features (heavy metals, trace elements, pathogens, etc.) must be considered.

(b) For simplicity's sake it is necessary to draw up a list of analyses required to establish the agronomic features of sludges and to standardise methods of analysis throughout the Community. This might take the form of a European standard which may perhaps later be incorporated in a Directive.

(c) Two precautions must be taken in using sewage sludge in particular :

(i)   as far as the sludge itself is concerned, precautions must be taken in its storage, transport and treatment, to control its odour, any pathogens or trace elements it may contain, the uses to which it is put, etc.

(ii) precautions must also be taken to protect the soil, i.e. to avoid the accumulation of undesirable elements in it.

This paper will discuss compost, effluent from stock rearing and sewage sludge in turn and describe the situation in each Member State.

Compost

The quantity of compost produced in the Community varies considerably from one Member State to another. Let us look at the situation in each country.

Federal Republic of Germany : approximately 2 per cent of domestic refuse, and almost the same amount of sewage sludge, is used for composting. There are nineteen plants, serving 1.9 million people and producing an estimated 130 000 to 150 000 tonnes of compost, i.e. 10 kg per person.

Netherlands : the Netherlands is the outstanding Community country having had a large compost industry since 1929. There are at present four plants which treat 450 000 tonnes of domestic refuse (i.e. 18 per cent of the total), but very little sewage sludge is composted. They produce a total of 70 000 tonnes of compost.

France : The composting industry is located mainly in the Paris region where 42 (out of a total of 86) composting plants serving 4.58 million inhabitants who produce 1.12 million tonnes of domestic refuse each year. An estimated 500 000 tonnes of compost is produced of which 17 per cent is used by mushroom growers, 61 per cent by vine growers and 10 per cent by vegetable and fruit growers.

United Kingdom : There is only one plant of any size, at Leicester, which treats 45 000 tonnes of domestic refuse each year. There are six smaller plants including four in Scotland.

Denmark : There is virtually no composting industry in Denmark. There are only two crushing plants which can produce compost, and only by the heap method.

Italy : Compost production represents 2.4 per cent of total domestic
refuse production (365 000 tonnes per year). The compost produced by
twelve composting plants is used mainly in horticulture and flower-
growing.

Belgium : There is one large composting plant in Gent which serves
160 000 inhabitants producing some 33 000 tonnes of domestic refuse.
A total of 20 000 tonnes of raw compost is produced each year.

Legal aspects of composting in Europe

There is no Community legislation on composting in the Community. A
comparison of national legislation reveals that :

   a)   the laws on composting vary widely from one country to
          another ;
   b)   not every country has national legislation in the usual
          sense of the term. In several Member States the intermediate
          and lower administrative levels have legislative power.

Federal Republic of Germany : Legislation applicable to composting is
made up of a number of separate laws, orders and implementing orders,
with Federal, land and local law superimposed upon each other

It covers the following :

   a)   the obligation to declare and monitor ;
   b)   collection, transport and composting of compostable waste ;
   c)   use of the final product of composting (compost) in agri-
          culture, arboniculture and horticulture.

United Kingdom : In contrast to the Federal Republic's system of
separate laws the United Kingdom has extended the scope of the Control
of Pollution Act to make it a comprehensive law. Compost and composting
are mentioned even less often in British than in German legislation.

France : Although composting is encouraged by the Government there is
only one, comprehensive law on water and environmental protection. The
legislative system in France is central (with no successive tiers of
competence as in the Federal Republic of Germany). The experimental
standard U44-051 of December 1974 which defines the raw materials for
the P, N, K and Mg content of composts is worth noting.

Italy : There is no national environmental legislation applicable to composting. The only valid rules on the construction and operation of composting plants are at regional and local level.

Denmark : There is a framework law on environmental protection (Law no. 372 of 13 June 1973) but this does not apply specifically to composting.

Conclusions

There are very great differences between national laws. At one end of the scale is Italy which has virtually no environmental laws on composting and at the other is Germany, where legislators are enacting special laws and orders to regulate an increasing number of processes techniques and situations. The United Kingdom and France are situated in the middle of the scale although they are far from comparable.

A comparison of the legal provisions applicable to composting in the Member States reveals that there are no specific regulations on composting.

The very fact that there are so many laws on composting within national environmental laws will hinder the creation of Community legislation. There is also little trade in this product.

Effluent from Stock Rearing

A study on the spreading of animal excrement on utilised agricultural areas of the Community has been published in the series "Information on Agriculture". The study's basic principles and recommendations can be summarised as follows :

(a) No more manure should be spread on the land then is necessary to obtain maximum crop production with acceptable quality. Surplus manure should be removed.

(b) It is recommended that the maximum quantities of manure used be determined according to the nitrogen requirements in the case of arable land and the potassium requirement in that of grassland. The acceptable quantity of manure and size of any possible surplus can be determined from the stocking rate, which is based on the assumption that one livestock unit (LU) is equivalent to the quantity of slurry produced annually by an adult cow containing 90 kg N, 40 kg P_2O_5 and 100 kg K_2O. The minerals contained in the manure of other types of animal can be expressed in cattle equivalent (CE).

(c) It is recommended that the copper content of pig feed be kept to a strict minimum and that soil analyses be made from time to time in fields on which pig manure is regularly spread. Pig manure should not be used on sheep grazing land.

(d) In order to prevent the pollution of surface water by phosphates, it is recommended that the quantities of manure spread on land which has a run-off of 1.5 CE-P per hectare of utilised agricultural land be limited and that information should be provided about the optimum timing of manuring in the light of factors such as intensity of rainfall, vegetation cover, topography, type of soil, etc.

(e) It is recommended that pollution of shallow groundwater by nitrates should be prevented or reduced as far as possible by basing the manure dosage and manuring timing on the growing period, type of soil, use to which land is put and rainfall.

(f) The quantities of nutrients already put into the soil by chemical or organic fertilisers, etc., should be deducted when calculating the maximum doses of manure to be applied. In some cases, the standards for certain types of land will have to be adjusted when there is a change in weather conditions because the standards relate to average conditions.

(g) The statistics obtained give a good idea of the size and geographical distribution of manure surpluses in the Community, particularly as the Member States have provided information for strictly delimited areas. They do not, however, reveal any need for standard rules on the use and disposal of particular quantities of manure to be applied to all farms in an area with a surplus of manure.

It is recommended that all livestock farms in areas in which the stocking rate exceeds a threshold value be given an opportunity to offer their surpluses voluntarily to a central body, e.g. a manure bank. Only if this solution did not have the desired result would measures be considered to be applied fairly quickly for individual farms.

(h) There may be surpluses on farms even in areas in which the stocking rate does not exceed the threshold value laid down. It may be assumed that these farms can easily dispose of their surpluses in neighbouring areas. Only if this proves not to be the case will measures be recommended in line with the policy adopted.

(i) More opportunities should be provided for non-land-dependent farming in arable and horticultural regions than in grassland areas to avoid a concentration of this type of farming and to facilitate the exchange of surpluses of manure. The economic, social and legal consequences must of course be taken into account.

(j) It is recommended that more information should be provided about the use and storage of manure and studies undertaken into the mechanism of water and soil pollution.

Use of sludge

An enormous amount of sewage sludge is produced in the treatment of effluent. In the Community this is estimated to be some 6 million tonnes of dry matter or more than 230 million m3 of raw sludge each year.

Each inhabitant thus produces 800 kg of sludge a year, in addition to bio-degradable industrial sludge (from breweries and the food industry).

A large increase in the sewage sludge produced in the Community can be expected. By 1990 it may be 15 to 20 x 106 tonnes of dry matter per year (according to ERL report, 1979).

Table I is a survey of sludge production and the disposal methods used in the Member States. Sludge disposal can in average terms be broken down as follows :

- 45 per cent of sludge is dumped on landfill sites
- 19 per cent is dumped at sea
- 7 per cent is incinerated
- 29 per cent is used in agriculture.

It is clear that the first three practices are no solution to the problem of sludge disposal ; indeed, dumping at sea or on land creates odour problems, eutrophication, and bacteriological and chemical pollution of water. Even if the sludge is incinerated, water and air pollution may still occur ; furthermore, the process is costly.

The 29 per cent of sludge used in agriculture is low considering that sludge is of definite value to the soil in that it is a source of organic matters, phosphorus and nitrogen. We shall see later why there

Sewage sludge : Production and utilization in the EEC   Table I

Country	Total production (i)	Future production	Utilization in agriculture	Incineration	Controlled dumping	Discharge at sea
Ireland	18.000 t/a (1977)	40.000 t/a (1977)	4 %		39 %	47 %
Germany	2.014.324 t/a (1974)	4.000.000 t/a (1985/1990)	34 %	8 %	51,6 %	?
France	700.000 t/a (1976)	increase of 5 a 8 % year	23 %	20 %	53 %	-
Belgium	56.000 t/a (1978)	increase of 28.000 t/a	10 %	10 %	80 %	-
U.K.	1.400.000 t/a (1975)	slight increase	44 %	3 %	33 %	23 %
Denmark	130.000 t/a (1972)	320.000 t/a (1982)	45 %	9,5 %	55 %	banned
Italy						
The Netherlands	137.000 t/a (1974)	400.000 t/a (1985)	31 %	2,6 %	4,9 %	
Luxembourg	6.200 t/a (1977)					

(i) : Total gross production (non-stabilized) expressed as dry matter per year
(ii): 33 % spread on non-agricultural land

is this lack of interest and what measures the Commission proposes to take to regulate the use of sludge in agriculture.

There is little legislation on sludge; if it exists at all it is part of framework laws or in its infancy.

In Germany, there is a Law on Waste Disposal (ABFG--1977), including an order on the discharge of waste water and sewage sludge (Article 15). There are also information notes (waste management) which recommend monitoring of the spreading of sewage sludge on agricultural land from both the hygiene and enviromental protection points of view (responsibil-- ity is shared between the 'Bund' and the 'Länder'). Finally, there are also information notes on the dumping of sewage sludge.

In France, the Law on the Protection of the Environment is slanted parti- cularly to the protection of water (one single law for the whole country). Even so, there is an experimental standard (AFNOR 44-041) on the classi- fication and specification of sewage sludges which lays down limits on the amounts of heavy metals which may be present in sludge.

In Belgium, there are specific regulations or recommendations concerning the use of sludge. It is covered by the Law on Pesticides and Raw Materials for Use in Agriculture, Horticulture, Forestry and Stockrearing of 11 July 1969 (Moniteur Belge of 17 July 1969) and the Law on the Protection of Surface Waters against Pollution of 26 March 1971 (Moniteur Belge of 1 May 1971).

The United Kingdom's Control of Pollution Act 1974 covers the dumping of waste on land and at sea and the incineration of waste. As for the use of sludge in agriculture, there is a Code of Practice (Standing Technical Committee Reports) which gives guidance, inter alia, on the maximum amount of heavy metals which may be applied to the soil.

In Denmark, there is only an 'outline law" on the protection of the environment (Law No 372 of 13 June 1973) but there are also recommenda- tions issued by the Danish Environmental Protection Agency concerning sludge from sewage works (report of a study group, February 1975); these recommend the types of treatment (stabilization - disinfection) and limit the amounts of heavy metals in sludge.

In Italy, Ireland and Luxembourg there are no specific regulations on recommendations on sludge.

In the Netherlands, there is a general law on waste (1977, stbH 455) and on fertilizers (1974, stbH 123) which are backed by royal decrees (1977). As far as sludge is concerned, legislation is being drafted on the concentration of heavy metals.

The use of sludge in agriculture poses serious problems as far as its treatment and the quantities spread are concerned, for both producers and users.

Sludge treatment

(a) Raw sludge is not used in agriculture for the following reasons :

    (i)   It is not stable and its uncontrolled fermentation releases foul odours;

    (ii)   The percentage of suspended solids is low (less than 2 per cent); use of such sludge would therefore involve the transportation of vast quantities of water for the very small quantities of substances of value to plant growth;

    (iii) The sludge has a high ratio of assimilable C to assimilable N. The soil micro-organisms would therefore take up the nitrogen – to the detriment of the plants – since the amount of nitrogen supplied by the sludge is inadequate;

    (iv) Raw sludge is a source of pathogens which may contaminate crops; this must under all circumstances be prevented, particularly in the case of crops for animal and human consumption.

(b) Various stages are involved in the treatment of sewage sludge,i.e.

    1.   Gravity or mechanical thickening which concentrates the liquid sludge coming from the settling tanks.

    2.   Biological or chemical stabilization which partly destroys the fermentable organic matter and retards the biological activity in the sludge.

    3.   Chemical or thermal conditioning to facilitate the subsequent dewatering stage.

    4.   Disinfection to cut down epidemiological hazards due to the sludge.

Quantities of sludge which may be spread

The determing factors are nutrient value and the level of trace elements in the sludge.

(a) Nutrient value

The nutrient content of sludge varies sharply depending on its origin and above all on the treatment it has undergone. In terms of dry matter, dewatered sludge has a lower nutrient value than liquid sludge since virtually all the potassium and soluble nitrogen compounds are removed along with the water.

Nitrogen values are between 1 per cent and 6 per cent, phosphorus (P_2O_5) between 3 per cent and 8 per cent, potassium 0.5 per cent and 1 per cent and organic matter 30 per cent and 80 per cent.

Consequently, since nitrogen and phosphorus are usually the elements most abundantly avaliable in sludge, they can be used to establish the rates at which sludge may be spread (the actual amount applied should be determined according to the level of trace elements).

(b) Level of trace elments

Sewage sludge generally contains very small quantities of metals and other chemical elements. But, if water reclamation plants also treat industrial effluent the sludge may contain toxic substances.

Excessive addition of trace elements to the soil may cause a number of things to happen, namely :
- a build-up in the soil, with potential cumulative toxicity;
- phytotoxic effects on those plants which absorb these elements;
- leaching and pollution of groundwater and surface water;
- accumulation and concentration throughout the food chain with the risk of toxicity for both animal and human consumers.

This is why it is absolutely essential to limit the quantity of trace elements supplied to the soil by sludge spreading. Some Member States have already adopted limit values for some trace elements (see table). These, however, vary considerably from one country to another and are expressed differently, making comparisons difficult.

In view of the problems involved in using sludge in agriculture the Commission is working on a draft Community Directive to not only

limit its use but also promote more efficient utilization. The
principles of the proposal Directive are :

I. Treatment

Raw sludge may not be used in agriculture. It must be stabilized,
whatever the crop, and must be disinfected for crops intended for
direct animal or human consumption.

II. Quantities to be spread

The maximum quantity of sludge which may be spread will depend on
the following criteria :

(a) the nitrogen and phosphate concentration in the sludge and
    in the soil;

(b) the type of crop;

(c) the concentration of trace elements in the sludge;

The quantity of N and P_2O_5 must not exceed 200 units per hectare
per year.

Sludge above the maximum concentration of trace elements may not
be used in agriculture.

Member States may lay down the maximum annual amounts of trace
elements per hectare but these values must be within the limits
proposed by the Directive. They must also take steps to ensure
that the maximum permitted concentrations in the soil are observed.

III. Sludge must be analysed prior to use as must soils to which
     sludge is applied and sludge - soil mixtures.

It is impossible to standardize methods of analysis in the
immediate future but the Member States will adopt common methods
of sampling, analysis and expression of results at a later
stage.

These analyses will be carried out by the persons responsible
for treating the sludge who must furnish users with the results.

IV. A number of recommendations on the spreading of sludge will also
    be put forward.

Utilization of sludge in agriculture : Limit values for trace elements (i) in Community countries   Table

Pays	Units	Zn	Cu	Mn	Pb	Cr	Ni	Co	Cd	Hg	Mo	B	Se	As
Ireland	non-existant													
Germany	mg/kg of dry sludge (ii) in sludge	2000 to 3000	400 to 600	500 to 1000	400 to 800				10 to 25	10 to 25	15 to 25			
France	mg/kg dry matter in sludge	3000	1500	500	300	200	100	20	15	8				
Belgium	non-existant													
U.K.	weight in kg/ha not to be exceeded in 30 years and more	560	280		1000	1000	70		5	2	5	5	5	10
Denmark	(a) : mg/kg of dry matter (b) : mg/ha/yr	6000 (a) 6000 (a)	700 (a) 700 (a)		600 (b) 1200 (b)	500 (a) 500 (a)			15 (b) 30 (b)					
The Netherlands	mg/kg of dry matter	2000	500		500	500	50		10	10				
Luxembourg	non-existant													
Italy	data not available													

(i) : Trace elements: chiefly the heavy metals (apart from Mo, B, Se, As) (ii) : mg/kg dry matter is equivalent to ppm in dry matter (iii) : 1st line-values for human and animal food crops 2nd line-values for non-food crops

RESEARCH ON THE UTILISATION OF SEWAGE SLUDGE

IN THE EUROPEAN COMMUNITIES

Dr H. Ott and P. l'Hermite

Directorate General for
Research, Science and
Education

Commission of the European Communities

SYNOPSIS

Since 1972, research on sewage sludge is coordinated on European level within the framework of a concerted action known under COST project 68. In this concerted action participate, in addition to the Community countries, also Austria, Switzerland, Norway, Sweden and Finland.

At present, about 400 research projects are coordinated.

The programme covers 5 main research areas :

- *the treatment of raw sludge in view of its use, in particular conditioning, dewatering and compostation ;*
- *the characteristics of sludge and the analysis of contaminants (heavy metals, organic pollutants) ;*
- *the identification and quantitative assessment of pathogens (bacteria, virus, parasites) and the effects of disinfection processes on the viability of pathogens in sludge ;*
- *the constraints to agricultural use of sewage sludge due to pollutant, in particular heavy metals (enrichment in soil, transfer from soil to crops, etc) ;*
- *the beneficial effects of sludge in agriculture (fertilizing value, improvement of soil quality) and methods for optimum application.*

The results achieved so far will be summarized, and the experiences made with the co-ordination of research on European level by a concerted action will be evaluated.

RESUME

*Depuis 1972, les travaux de recherche sur les boues d'épuration sont
coordonnés au niveau européen dans le cadre d'une action concertée
dénommée "Projet Cost 68". Aux pays de la Communauté qui participent à
cette action concertée s'ajoutent l'Autriche, la Suisse, la Norvège,
la Suède et la Finlande. A l'heure actuelle, quelque 400 projets de
recherche sont coordonnés.*

Le programme couvre cinq grand domaines de recherche :

- *Le traitement de boues à l'état brut en vue de les utiliser notamment
  le conditionnement, la déshydratation et la détermination de la
  composition.*

- *La caractérisation des boues et l'analyse des contaminants (métaux
  lourds, polluants organiques).*

- *L'identification et l'évaluation quantitative des agents pathogènes
  (bactéries, virus, parasites) et l'incidence des procédés de désinfec-
  tion sur la viabilité des agents pathogènes dans les boues.*

- *Les obstacles à l'utilisation agricole des boues d'épuration dus aux
  polluants, notamment aux métaux lourds (enrichissement du sol, trans-
  fert du sol aux cultures, etc.)*

- *Les essais bénéfiques des boues en agriculture (pouvoir fertilisant,
  amélioration de la qualité du sol) et les méthodes d'utilisation
  optimale.*

*On rédigera un résumé des résultats obtenus jusqu'ici et l'on évaluera
les expériences en matières de coordination des travaux de recherche au
niveau européen dans le cadre d'une action concertée.*

ZUSAMMENFASSUNG

*Seit 1972 werden die Forschungsarbeiten über Klarschlamm auf europäischer
Ebene im Rahmen einer konzertierten Aktion, der "Cost-Aktion 68",
koordiniert. Neben den Mitgliedstaaten der Gemeinschaft sind an dieser
konzertierten Aktion auch Österreich, die Schweiz, Norwegen, Schweden und
Finnland beteiligt. Derzeit werden rund 400 Forschungsvorhaben koordiniert.*

Das Programm umfasst im wesentlichen 5 Forschungsbegiete :

- Rohschlammbehandlung vor seiner Verwerdung, insbesondere Stabilisierung,
  Entwässerung und Zusammensetzung.

- Charakterisierung von Klärschlamm und Schmutzstoffanalyse (Schwermetalle,
  organische Schandstoffe).

- Feststellung und Mengenbestimmung pathogener Organismen (Bakterien, Viren,
  Parasiten) sowie Auswirkungen von Desinfektionsverfahren auf die Leben
sfähigkeit pathogener Organismen im Klärschlamm.

When, in the early seventies, the European Communities started to set up an Environmental Research Programme, sewage sludge was among the problems to be tackled first.

Research on sewage sludge started within the framework of COST (Cooperation Scientifique et Technique). Nineteen European countries came together at the initiative of the European Communities to discuss cooperation in the scientific and technical field. This initiative led to the definition of a number of research projects which should be coordinated on European level, among them one entitled "Sewage Sludge Processing" (COST Project 68). The multilateral agreement on the coordination of research in thirteen countries (Netherlands, Italy, Ireland, Denmark, United Kingdom, Sweden, Finland, Norway, Belgium, France, Austria, Switzerland and Germany) as concluded in 1971, the project was finally launched in late 1972, and was concluded in 1975. The main results are laid down in a Final Report [1].

The satisfactory outcome of this project encouraged the Commission of the European Communities to take the initiative of launching a follow-up project. In 1977, the Council of the European Communities decided a research programme "Treatment and Use of Sewage Sludge", to be implemented as " Concerted Action" for a period of three years [2]. In 1978, an Agreement within the framework of COST between the European Communities and five European Non-Member States (Austria, Switzerland, Norway, Sweden and Finland) was signed, which constitutes the legal basis for the association of these countries to the Community concerted action, now known as COST project 68 bis [3].

A further extension of the research programme on sewage sludge unitl 1983 is at present under preparation.

The purpose of this text is to :
- summarize the main results of COST project 68 ;
- outline the programme of the currently ongoing concerted action (COST project 68 bis) and to review the results achieved so far.

Furthermore, the general principles of Concerted Actions and their role within the research activities of the European Communities will be explained.

Results of Cost Project 68 "Sewage Sludge Processing"

The aims of the COST Project 68 "Sewage Sludge Processing", implemented

from 1972 to 1975, were limited to some selected aspects of sewage sludge
treatment and disposal.

The three main topics were :

    a)   standardization of characteristic values of sludge :

    b)   elaboration of methods for the characterization of sludge
        parameters ;

    c)   comparative tests of combined sludge-refuse incinerators.

The final report [1] published in 1975 contains a list of definitions
for sludge parameters and methods for their determination adopted for
the European Cooperation.

Two methods, jointly elaborated and tested, were described and proposed
as "recommended standard methods" : the determination of "conditionability"
and the assessment of particle size distribution. At least the first
mentioned method found in the meantime a wide field of application.

The main work was, however, the comprehensive testing of the performance
of two installations for combined incineration of sludge and refuse, in
which mass and energy balances were established and pollutant emissions
determined.

This work fell together with a rapidly growing prize for primary energy,
and incineration of sludge became less and less attractive. Work on
incineration was therefore discontinued at the end of COST project 68.

<u>The Concerted Action "Treatment and Use of Sewage Sludge" (COST 68 bis)</u>

a)   <u>The Aims of the Programme</u>

At the beginning of the planning for a new programme in 1975, it was
assumed that the drastically reduced economy of incineration and the
increasing price for mineral fertilizers would up-grade the "waste"
sewage sludge to a valuable resource. A careful analysis of the results
of the preceding programme led to the conclusion, that further research
should mainly aim at assessing and eliminating the constraints to
spreading sewage sludge on cultivated land.

There is a justified concern about the consequences of land spreading
which are essentially the following :

    -   sludge contains pollutants, in particular heavy metals, which
        may accumulate in the soil and contaminate the crops ;

- pathogens may be disseminated :
- excessive or inappropriate application may lead to ground water pollution :
- long term application might deteriorate soil quality :
- spreading may cause odour nuisance and esthetic problems.

Furthermore, it was clear that the economy of sludge treatment and application should be improved. Last but not least the programme should help to overcome major psychological problems with sludge spreading, involving farmers as well as general population.

In the Council decision on the programme, the following general framework was adopted :

a) Sludge stabilisation and odour problems :
- definition and determination of "degree of stability" and relationships to odour nuisance ;
- comparative evaluation of stabilisation procedures.

b) Problems related to sludge dewatering :
- research on water binding forces :
- development and standardization of methods for the assessment of dewatering properties ;
- problems related to the use of flocculants ;
- comparative evaluation of thickening and dewatering equipment.

c) Analytical problems related to sludge treatment and use :
- characterisation of pathogens and evaluation of disinfection procedures ;
- characterisation and determination of pollutants (heavy metals, persistent organic compounds in sludge and development of standardized analytical methods.

d) Environmental problems related to sludge use :
- special processing of sludge for agricultural use (e.g. composting) including the improvement of disinfection procedures and pollutant removal ;
- transfer of pollutants to plants and harmful effects on vegetation ;
- effects of long range sludge application on soil quality and ground water ;
- optimum land use of sludge.

In defining the new programme, emphasis was put on the elaboration and validation of methods which were supposed to be necessary for the implementation of regulations to be set up within the framework of the Environmental Action Programme of the European Communities.

b)   The Principles of Concerted Actions

The Community has three different ways of promoting research :   direct action,   i.e. research executed intra muros in the Communities own Research Centre, indirect action, implemented by research contracts with competent institutes in the Communities, and concerted action, which is defined as coordination at Community level of all relevant nationally funded research in a given field.

Mainly for historical reasons, research on sewage sludge is implemented exclusively as concerted action.

The term "coordination" in this context stands mainly as for mechanism for the efficient exchange of information;   no attempt is made to influence directly the research content of national projects.

The Commission's Services are assisted by a Committee composed of those responsible for the oordination of research on national level.   The concept of pilot countries, assisted by co-pilots, which took the responsibility for the coordination of working parties in different fields has been adopted.   These working parties report to the Committee which assumes the task of compiling and disseminating the results.

The organisation of symposia and workshops is another tool of coordination. Within the framework of COST Project 68 bis, the first Symposium was held in 1979 in Cadarache, the second is scheduled for late 1980 in Vienna.

c)   Implementation of the Programme

The first step in implementing the programme was the establishment of an inventory of the ongoing or planned research projects.   The inquiry resulted in an inventory of more than 400 projects, which served as a basis for the coordination effort.

Within the framework of COST project 68 bis, five working parties have been established;   their terms of reference are as follows :

1)   Sludge Processing
Technical aspects of sludge processing (stabilization, dewatering etc.)

and definition of parameters relevant to the processes (including analytical methods for their characterisation) :

2)   Chemical pollution of sludge
Analytical aspects of the determination of heavy metals and organic pollutants, characterisation and measurement of odours, chemical state of pollutants, decontamination, etc. ;

3)   Biological pollution of sludge
Characterisation of bacteria, virus, parasites and all aspects of disinfection ;

4)   Valorisation of sludge
Fertilizing properties and other positive aspects (e.g. improvement of soil structure), optimum application, improvement of product quality, etc.;

5)   Environmental effects of sludge
Potential constraints to agricultural use, due to heavy metals, organic pollutants, and pathogens; contamination of soil, ground and surface water; transfer of pollutants to crops.

It is not the scope of this paper to give details of the scientific results, which are laid down in a number of reports, at present being processed for publication. In the following, its emphasis is therefore given to the systematic approach which was made.

Working party "Sludge processing" prepared first a flow chart on sludge treatment and disposal, facilitating the identification of process steps possibly being considered for research initiatives, and established priorities within them. The main aims are the improvement of the economy of the processed and the reduction of the polluting potential of the products.

Working party "Chemical Pollution of Sludge", established a list of heavy metals for priority consideration in view of a standardization of harmonisation of analytical procedures for their determination in sludge, including methods for sludge mineralisation. Organic pollutants will be considered at a later stage. A number of defined samples were circulated for comparative analysis. The results of the first campaign were presented at the Cadarache Symposium and carefully analysed. Further intercomparison campaigns are at present being evaluated or in preparation.

Working Party "Biological pollution of sludge" generated an inventory of microorganisms and parasites of importance and defining priorities for their investigation, in particular with regard to determination methods. Emphasis is put now on exploring which microorganisms (or group of them) could serve as indicator for the evaluation of disinfection procedures.

Working Party "Valorisation of sludge" generated a comprehensive report reviewing in detail the situation of sludge use on agricultural land in the participating countries. The available knowledge on the availability of nitrogen of sludge was subject of a workshop and will be published later this year. Similar work with regard to phosphorous is going on.

Working Party "Environmental effects on sludge" has concentrated its attention on the effects of heavy metals and pollution of ground water. Guidelines for the execution of field experiments have been established as an essential prerequisite for making results in a concerted action comparable. A report on Cadmium in sludge, considered to be the most important pollutant, is in preparation.

It should be mentioned that all effort is made to coordinate work on sewage sludge with the agricultural research programme of the European Communities, in particular with its part on effluents of livestock.

Future Developments

The concentrated effort on compiling the results of European research on sewage sludge gave a rather impressive picture with regard to the already available knowledge in some of the areas; other fields, however, in particular microbial pollution and effects of long term application to land, seem far from being well understood. There still seem to be possibilities also for the improvement of sludge treatment and application methods. All scientists involved advocated strongly for a continuation of the concerted action. The Commission proposed therefore to the Council an extension of the programme until 1983 as a part of the sectorial environment research programme.

References :

1) COST Project 68, Sewage Sludge Processing, Final Report of the Management Committee (Doc. EUCO/SP/48/75), 1975, available at request from the Commission of the European Communities.

2)   O.J. no. L 267/35 of 19.10.77.

3)   O.J. no. L 72/35 of 23.03.79.

QUESTIONS & COMMENTS

Answering a question put by Mr Richard Baker, of the UK Conservation
Society, Dr Ott said that there were problems in the use of effluent from
intensive livestock farming, notably pig rearing, because the material was
so concentrated that it could contaminate ground water.

Mr Vellaud commented that industrial waste products that could be used on
the land included the residue of vitamin B 12 which was rich in nitrogen
and the sludge of decarbonating process water which could be used as a pH
ameliorator for very acid soil.

Answering a question on heavy metal contamination, Dr Ott pointed out the
need for a fairly simple analysis method for sludge for use in agriculture.
This would avoid the situation where the cost of analysis exceeded the
value of the sludge as a fertiliser.

THE USE OF WASTE IN AGRICULTURE

Jean-Paul Vellaud

National Waste Disposal and

Recovery Agency

France

SYNOPSIS

The need to protect the environment and conserve raw materials is a direct incentive to make greater use of organic waste in agriculture.

The increasing number of treatment plants, industry and towns are faced with a new problem : disposing of the sludge from water-treatment processes. Some 100 million tonnes of wet sludge (5.5 million tonnes of dry matter) are produced each year in the Community. Sludge contains nutrients, particularly nitrogen, phosphorus and organic matter, which are essential to agricultural land; it may also contain metals and pathogens which are hostile to agricultural activities. The use of sludge of known composition in agriculture should be encouraged; it helps farmers to make substantial savings in the fertilizers and organic matter which they need for their crops.

The system of sludge spreading which has been operating for some years in the Federal Republic of Germany is evidence of how effectively such a system can be organized and the services it can provide to local authorities and farmers.

The advice given to farmers in France on the use of sludge in agriculture has helped to develop a system of direct utilization on a local basis.

The use of domestic refuse as a soil conditioner is not new. However, it is only in the last ten or so years that there has been controlled production of good quality urban compost; the Community produces only 700 000 tonnes of compost a year. The principal value of the compost is in the organic matter which it contains; as agronomists have indicated, a shortage of organic matter in the soil leads to low crop yields in particular. Compost may be "soiled" by foreign bodies such as glass and plastics which limits the use to which it can be put.

Holland has set up a unique national system of domestic refuse collection and composting which has given Dutch and foreign farmers a wide range of valued products.

Of all industrialized countries France has the largest number of composting plants. Because the compost produced has not always been of good quality the French authorities have set up an advisory service for farmers to help make the unique potential of plants in operation more efficient and produce compost of a quality required by users.

RESUME

La nécessité de préserver l'environnement et de réduire le gaspillage de matières premières constitue une invitation vigoureuse à développer la valorisation des déchets organiques en agriculture.

La multiplication des stations d'épuration place les industriels et les villes devant un problème nouveau qui est celui de l'élimination des boues produites par le processus de traitement des eaux. C'est environ 100 millions de tonnes de boues humides, soit 5,5 millions de tonnes de matières sèches qui sont produites chaque année dans la CEE. Ces boues contiennent des éléments fertilisants ; l'azote et le phosphore notamment et de la matière organique dont les sols agricoles ont besoin. Elles peuvent contenir également des éléments métalliques et des germes pathogè-nes dont l'agriculteur se méfie. L'utilisation en agriculture de boues dont la composition est connue, contribue au niveau de l'exploitation agricole, à des économies substantielles d'achat d'éléments fertilisants et de matière organique nécessaires à l'exploitation. Elle doit donc être développée.

L'exemple allemand qui s'appuie sur de nombreuses années d'expérience, témoigne de l'efficacité d'une organisation mise en place pour l'épandage des boues et des services qu'elle apporte aux communes et aux agriculteurs.

En France, le conseil apporté aux agriculteurs sur l'utilisation agricole des boues permet de développer des circuits courts de valorisation des boues bien adaptés à la structure du pays.

L'utilisation des déchets ménagers pour amender le sol est une pratique ancienne. Cependant la fabrication controlée d'un compost urbain de bonne qualité n'a été maitrisée que depuis quelques dizaines d'années et la CEE ne produit que 700.000 tonnes environ de compost par an. Le compost est surtout intéressant par la matière organique qu'il contient dont le déficit dans le sol est mis en évidence par les agronomes et entraîne notamment des diminutions de rendement. Le compost peut être souillé par des corps étrangers - verre, plastique - ce qui freine son emploi.

La Hollande a mis en place au niveau national une organisation originale de collecte des ordures ménagères et de fabrication de compost qui permet d'obtenir des produits très variés et appréciés par les agriculteurs hollandais et étrangers.

La France dispose du plus grand parc d'usines de compostage en activité des pays industrialisés. Le compost produit n'a pas toujours été de bonne qualité, cela conduit les autorités françaises à mettre en place une structure de conseil aux exploitants pour contribuer à mieux faire fonc-tionner un potentiel d'installations uniques et fabriquer du compost qui réponde à la demande des utilisateurs.

ZUSAMMENFASSUNG

Die notwendigkeit, die Umwelt zu schützen und die Verschwendung von Rohstoffen einzuschränken, gebietet es, Methoden zur Nutzung organischer Abfälle in der Landwirtschaft zu entwickeln. Ver mehrt gebaute Kläranlagen stellen Industrie und Städte vor das neue Problem, den in dem Klärverfahren anfallenden Schlamm beseitigen zu müssen. Innerhalb der EG werden jährlich etwa 100 Mio Tonnen an feuchtem Schlamm produziert; das entspricht 5.5 Mio Tonnen Trockensubstanz. Dieser Schlamm enthält anorganische Stoffe, insebesondere Stickstoff und Phosphor, die zue

Düngung verwendet werden können. Ausserdem enthält er organische Substanzen, die der Boden braucht. Er kann jedoch auch Metalle und Krankheitserreger enthalten, vor denen die Landwirtschaft sich hütet. Die landwirtschaftliche Nutzung eines Schlamms, dessen Zusammensetzung bekannt ist, trägt zu bedeutenden Einsparungen bei Düngemitteln und notwendigen organischen Stoffen bei, und muss weiter entwickelt werden.

Das deutsche Beispiel, das sich auf die Erfahrungen aus vielen Jahren stützt, beweist die Leistungsfähigkeit einer Organisation, die für die Verteilung des Schlamms auf die Flächen sorgt, sowie die Leistungs-fähigkeit ihrer Dienstleistungen die sie Gemeinden und Landwirten zur Verfügung stellt.

In Frankreich werden die Landwirte bei der Verwendung des Klärschlamms beraten; dadurch kann sich eine direkte Nutzung von gut an die Gegenbenheiten angepasstem Schlamm entwickeln.

Die Verwendung von Haushaltsabfällen zur Bodenverbesserung ist ein altes Verfahren. Jedoch erst seit gut 10 Jahren beherrscht man die kontrollierte Fabrikation von gutem, urbanem Kompost. In der EG werden jährlich nur etwa 700 000 Tonnen Kompost produziert. Der Kompost ist for allem aufgrund der darin enthaltennen organischen Stoffe interessant; denn von den Agronomen ist gerade ein Mangel an organischen Stoffen im Boden festgestellt worden, und ein solcher Mangel führt insbesondere zu geringeren Ausbeuten. Der Kompost kann jedoch z.B. durch Glas und Kunstoff verunreinigt und damit vermindert brauchbar sein.

Die Niederlande haben auf nationaler Ebene eine Organisation eintwickelt, die Haushaltsabfälle sammelt und Kompost produziert. Es können auf diese Weise sehr unterschiedliche Produkte hergestellt werden, die von den holländischen und ausländischen Landwirten sehr geschätzt werden.

In den Industrieländern verfügt Frankreich über die meisten Kompostierwerke. Der dort hergestellte Kompost war nicht immer von guter Qualität. Das hat die Französischen Behörden veranlasst, Beratungsstellen einzurichten. Auf diese Weise tragen sie dazu bei, das einmalige Potential an Fabrikationsstätten besser zu nutzen, und Kompost herzustellen, der den Bedürfnissen der Verbraucher besser entspricht.

High prices of raw materials and export deficits coupled with environmental considerations, provide a strong incentive to reduce raw materials
wastage.

Wastes composed primarily of organic matter include these from agriculture, sewage sludge and municipal refuse primarily from households.
The farming world has long recognised the importance of re-using manure
as a solution to its disposal problem. However, the nature and the
quantities of agricultural wastes offered to farmers have greatly changed
in the last few years and this has - rightly - put users on their guard.

Although the agricultural re-use is nothing new in the Member States,
official interest by Community institutions started only in 1979, when a
Working Party was set up to look into the matter.

Sludge, which is made up of suspended matter and micro-organisms, is
produced at the rate of 2.5 litres per person per day. Dry matter content
is 50 g. In other words, a water reclamation plant serving a population of
100 000 produces 250 m^3 of sludge containing 98 per cent water per day.

Sludge composition varies according to the characteristics of the effluent
being treated and the type of treatment used. Table I illustrates the
average composition of urban sludge compared with liquid manure and dung.

When added to the land, the organic substances in sludge decompose into
stable organic matter "jumus" which improves the structure and stability
of the soil and makes it more workable.

There are three main mineral elements also present in sludge, the most
important of which is nitrogen, since it affects crop yield. Nitrogen in
sludge is present as a mineral directly assimilable by plants, but is also
present in organic form (50 - 90 per cent of total nitrogen) and cannot
be used by plants until it has undergone a gradual transformation process
by the microflora present in the soil.

Consequently it is not enough to know the total concentration of nitrogen
in a given sludge in order to assess its value as a nitrogenous fertilizer.
Furthermore, nitrogen not used by crops can pollute ground water.

A significant proportion of the phosphorus present in sludge - at least
70 per cent - is present in mineral form and therefore assimilable by
plants. Phosphorus is readily fixed by the soil and its presence alone
may - in certain cases - justify the spreading of sludge.

Table I

Average composition of municipal sludge

A.N.R.E.D.

Principal components	Dry matter	Organic matter	Total nitrogen	Phosphoric acid	Potassium
Semi-liquid manure	10 %	600à800(1) 60à 80(2)	40 à 60 4 à 6	20 à 50 2 à 5	30 à 50 3 à 5
Farm manure	25 %	600(1) 75(3)	10 à 30 1,25à3,75	3 à 25 0,4à3,15	25 à 35 3,15à4 4
Aerobically-stabilized urban effluent	6 %	450à600(1) 27à 36(2)	45 à 60 2,7à3,6	40 à 85 2,4à5.1	5 à 15 0,3à0,9

Secondary components	Dry matter	Ca	Mg	Fe	Mn	B	Zn	Cu
Aerobically stabilized urban effluent	6 %	1 à 250(1) 0,6à 15(2)	2 à 13 1,2à0,8	5 à 15 3 à 9	0,1à0,5 0,07à0,3	0,2à1,5 0,12à0,9	1 à 3 0,6à1,8	0,1 à 0,06à

(1) kg/t dry matter
(2) kg/m3 of product
(3) kg/m3 of manure

Potassium is scarcely affected by the water treatment process and is therefore present in small concentrations in the sludge. Calcium may also be present in limed-sludges which may be regarded as a calcium soil amendment; magnesium may also be present.

Sludge usually contains low concentrations of other chemical elements of industrial origin, some of which are vital to crop growth up to certain concentrations, but are toxic if excessive. These trace elements comprise zinc, copper, manganese, boron, molybdenum, cobalt and iron. Other minerals, such as lead, mercury and cadmium, are of no value and above certain concentrations contaminate soil and plants.

Certain sludges may additionally contain viruses and pathogens. Treatment should centre on the conditions of use (e.g. wait two months before putting animals out to grass on fields treated with suspect sludge) rather than on disinfection of the sludge.

In 1979 the theoretical value of sludge, bearing in mind its contents and expressed per tonne of dry matter ranged from FF 150 to FF 250 in France. But its commercial value was only FF 50 - 100.

Use in agriculture

The annual production of sludge (expressed as dry matter) in the Member States is approximately 5.5 million tonnes (see Table II) - equivalent to just over 100 million tonnes of liquid sludge.

Germany tops the list with a production of just over 2 million tonnes, whereas the annual arising in Luxembourg is only 6 200 t.

In those Member States where statistics are available, it is abundantly clear that the use of sludge in agriculture is now a reality; the practice is most widespread in Holland, where almost 80 per cent of such sludge is spread on both agricultural and other types of land. Next come Denmark (45 per cent), United Kingdom (44 per cent), Germany (34 per cent) and France (23 per cent).

Regulations

The regulations adopted by individual Member States (Table III) to control the risks of a build-up of substances which might prove harmful for the soil, plants and water above certain concentrations relate to the following limits :

Table II

EEC production and utilization of sewage sludge

Country	Total Output (1)	Utilization in agriculture	Incineration	Controlled tipping	Dumping at sea
Germany	2.014.324 T/a (1974)	34 %	8 %	51,6 %	
France	1.700.000 T/a (1979)	23 % (2)	31 % (2)	46 % (2)	--
Italy		--	--	...	--
Netherlands	137.000 T/a (1974)	45% + 33% (3)	2% 20% (74) (80)	?	2% 0% (74) (80)
Belgium	56.000 T/a (1978)	10 %	10 %	90 %	--
Luxembourg	6.200 T/a (1977)	--	...	¦¦	..
United Kingdom	1.400.000 T/a (1975)	44 %	3 %	33 %	23 %
Ireland	18.000 T/a (1977)	3,5 %	--	52 %	44,5 %

(1) Total gross outpunt (non-stabilized) expressed as dry matter per year

(2) This breakdown only covers municipal sewage sludge: 760 000 t dry matter/yr

(3) 33 % spread on non-agricultural land

A.N.R.E.D.

Table III

Economic re-use of sludge in agriculture: Maximum limits on specific elements (1) in the countries of the EEC

Country	Units	Zn	Cu	Mn	Pb	Cr	Ni	Co	Cd	Hg	Mo	B	Se	As
Germany	mg/kg of dry matter (2) in the sludge	2000à 3000	400à 600	500à 1000	400à 800				10à 25	10à 25	15à 25			
France	mg/kg dry matter in the sludge	3000	1500	500	300	200	100	20	15	8				
Italy	data unavailable													
Netherlands	mg/kg dry matter	2000	500		500	500	50		10	10				
Belgium	not applicable													
Luxembourg	not applicable													
United Kingdom	"weight in kg/ha not to be exceeded over a period of 30 years or more	560	280		1000	1000	70		5	2	5	5	5	10
Ireland	not applicable													
Denmark (3)	a) mg/kg dry matter b) mg/ha/yr	6000 (a) 6000 (a)	700 (a) 700 (a)		600 (b) 1200 (b)	500 (a) 500 (a)			15 (b) 30 (b)					

(1) principally heavy metals with the exception of Mo, B, Se and As

(2) mg/kg dry matter is equivalent to ppm in the dry matter

(3) first line : values for crops for human and animal consumption;
    second line: values for non-food crops

Table IV

Composition of domestic refuse (as a percent of the total weight)

Constituents	Germany	France	Italy(Rome)
paper and board	28	20 – 35	20
glass	9	5 – 10	6
metals	7	5 – 8	3
plastics	3	3 – 6	4
textiles	3	1 – 5	–
organic wastes	15	15 – 35	39
fine waste	28	10 – 20	15
other residues	7	–	13

Table V

Compost production in the EEC

Country	Tonnage ('000 tonnes) of wet domestic refuse	Number of composting plants	% of domestic refuse treated by composting	Tonnage of composting manufactured per year
Germany	15.500	19	3 %	150.000
France	13.000	88	9,7%	420.000
Italy	13.700	–	–	–
Netherlands	3.360	4	18 %	70.000
Belgium	2.440	1	::	18.000
Luxembourg	88	–	–	–
United Kingdom	14.000	1	–	20.000
Ireland	760	–	–	–
Denmark	1.260	2	::	–

Sources : Mr Just (INRA 1977)

Mr Bernard (DCLP)

TABLE VI

CHEMICAL COMPOSITION OF COMPOST AND DOMESTIC REFUSE IN FRANCE

Water content per cent	35
pH	7.8

As per cent of the dry product

Carbon	15
Nitrogen	0.9
Phosphoric acid (P_2O_5)	0.6
Potassium (2kO)	0.3
Calcium	4
Magnesium	0.3
Sodium	0.3
Suplhur	0.6
Chlorine	0.5
Iron	0.2

In ppm of the dry product

Total boron (1)	64 - 245
Soluble boron (1)	10 - 31
Zinc	1 000
Manganese	600
Copper	250
Lead	590
Cadmium	7
Chromium	270
Nickel	190
Mercury	4

(1) According to Troome, Boniface 1977 - extreme values from 56 samples

- limit on the concentration of specific elements in sludges (France) ;
- limit on the total amount of specific elements added to the soil or spread on the land over a given period ;
- limit on the amounts applied in dressings which takes account of the relative toxicity of particular elements by using the "zinc equivalent" concept (United Kingdom).

In the light of all the findings from the research, it would be both possible and desirable to harmonize the regulations at EEC level.

Compost from domestic refuse

The first attempts at industrial-scale fermentation of domestic refuse in Europe were carried out in 1912 by Professor Beccari, notably at Avignon (France).

According to the latest definition, composting is "the crushing and screening of domestic residues followed by a carefully-controlled fermentation process in which organic substances, nitrogen, phosphorous and specific trace elements can be recovered with a view to being used in agriculture".

One of the major differences between sludge and compost is the heterogeneity of compost, whereas sludge is a liquid medium which is readily homogenized. This characteristic of compost derives from the composition of domestic refuse (see Table IV) and causes difficulties in taking representative samples of a batch of a particular compost.

The average composition of French compost is shown in Table VI. The composition varies according to the types of domestic residues treated and the season of the year. This table highlights the presence in compost of organic substances, mineral and chemical elements.

Compost is rich in organic matter, which accounts for some 35 per cent of the dry matter content. Under the same conditions the carbon content is approximately 15 per cent of the dry matter content.

The spreading of compost in normal doses (30-40 t/ha) on crops has a significant effect on the humus balance of the soil, which is more marked than the results obtained with sludge.

Compost derived from domestic refuse normally has a low concentration of mineral elements. Nitrogen accounts for less than 1 per cent of the dry matter and its behaviour in the soil depends on the level of maturity of the compost, generally expressed by the C/N ratio.

Fresh compost has a C/N ratio greater or equal to 30 and may even reach 100 or 120, in which case the micro-organisms present in the soil receiving such compost would extract the mineral nitrogen in the soil for their growth, this occasioning a nitrogen deficiency for the plants. On the other hand, mature compost, i.e. where the C/N ratio is of the order of 15 or 20, would have no inhibiting effect on plants.

The amounts of phosphorus and potassium present are low, namely 0.4 to

0.7 per cent for phosphorus and 0.4 to 0.2 per cent for potassium. On the other hand, compost generally contains 2.5 to 5 per cent of calcium, meaning that it can render acid soils more aklaline.

Although the chemical elements are present in generally low concentrations, care must nevertheless be taken to avoid their build-up in specific crops, especially mushrooms. However, problems are less than with sludge.

From a health point of view, the aerobic fermentation which the compost undergoes during preparation raises its temperature to 60 - 70°C, at which most pathogenic germs adapted to human and animal body temperatures are killed.

Finally, the quality of compost is largely dependent on its external appearance, as measured by foreign body content and particle size. Foreign bodies include glass, plastics and metallic objects, notably sharp-pointed ones. The hostility of farmers to glass is often unjustified because fragments will have lost their sharp edges.

Utilization of compost

From the statistics available, the countries of the EEC (Table V) produce some 700 000 tonnes of compost per year.

In the Federal Republic of Germany, 3 per cent of domestic refuse is treated by 19 plants going mainly to wine growers, but farmers are wary of substances present in toxic amounts, especially heavy metals. This has held back the construction of composting plants called for by a number of specialists.

In the United Kingdom, the only major composting plant processes half the
domestic refuse from Leicester, producing a high-quality compost sold
mainly in bag form. In Denmark, two composting plants operate but there
is strong opposition from farmers to using it, particularly as organic
needs can be met by manure.

In Belgium, the only composting plant of any size is located near Gent in
an intensive farming area; Italy has a few large composting plants
including one in Rome. Of the industrialized countries, France
undoubtedly has the longest experience of large-scale composting, producing some 420 000 tonnes per year.

Member States' regulations on composting deal with the problems either
from the industrial point of view, i.e. the plant is governed bv the
regulations applicable to industrial establishments which cause a nuisance,
or from an agronomic point of view, i.e. the compost used in agriculture
must comply with specific physical or chemical criteria.

In France, the operation of a composting plant is subject to an administrative authorization accompanied by certain requirements. Minimum
amounts of organic matter, total nitrogen, as well as particle sizes
have to be respected. In West Germany the waste disposal law applies,
but other countries of the EEC do not appear to have specific regulations.

One of the two major producers of compost, the Netherlands, had as its
original intention not waste disposal but to render sandy soils fertile.
Of the six composting plants operating in the mid-sixties, four are still
in operation, two at Wijster and Merlo (VAM), one at Ripenhout and one
serving the communes of Baarn and Soest. Many small plants producing a
lower-grade compost have disappeared either because of marketing difficulties or because of competition from the major incineration plants.

In the other major producer, France, a recent survey on the treatment of
domestic refuse showed that 88 plants were operating, serving a population
of 5 000 000, and producing 420 000 tonnes of compost per year. However,
marketing problems are encountered by the majority of plants, operated by
local authorities or private firms, arising principally from high foreign
body contents; mistrust among users ; and physical and chemical
properties not meeting user requirements. As a result, there has been a
gradual falling-off since 1970 in the percentage of the population
served by composting plants.

In order to encourage composting, the French Government – through the National Waste Disposal and Recovery Agency – is launching a program, including the provision of a touring mobile laboratory analyse commercial composts and give other advice. Development of high-quality brands of compost is also part of the campaign.

It must be stated in conclustion that although sludge and compost are already used in agriculture throughout the EEC, there are a number of inhibiting factors. Principal among these is the lack of information on the physical and chemcial properties of the sludge and compost offered to the farmer which puts him on his guard.

It should be possible to give manufacturers and users essential product information in order to remove this barrier. A vigorous campaign along these lines should be mounted.

ENERGY FROM MUNICIPAL WASTE
IN THE EUROPEAN COMMUNITY

Leopold Van Wambeke

Directorate-General for Research, Science and Education

Commission of the European Communities

SYNOPSIS

*Data are provided about municipal wastes arising in the European Community
(95 million tonnes) and about their composition. Conventional incineration
still remains the sole operational method for energy recovery but at
present only 13 per cent of the municipal waste is processed in such a way.
There is, however, a general trend to recover more energy owing to the
rising cost of energy raw materials and the scarcity of landfills. Some
plants are near the stage of profitability.*

*The environmental problems related to the incineration of municipal waste
are discussed, the main one being emissions into the atmosphere and
pollution which may be generated by the deposit of fly ash. They may be
sharply reduced by adequate equipment and safe operating measures.*

*The new techniques of energy recovery from municipal waste, still all in
the experimental stage, are also briefly described.*

*A series of guidelines and recommendations are given in the conclusions
as well as some R & D needs.*

RESUME

*Il existe des données sur les quantités de résidus urbains produits dans
la Communauté européenne (95 millions de tonnes) et sur leur composition.
L'incinération traditionnelle est encore la seule méthode appliquée pour
récupérer l'énergie , mais pour le moment, 13% seulement des résidus*

urbains sont traités de cette manière. La tendance générale veut
cependant que l'énergie soit de plus en plus récupérée en raison de
l'augmentation du coût des matières premières énergétiques et du manque
de décharges. Certaines installations ne sont pas loin du seuil de la
rentabilité.

On examine les problèmes écologiques liés à l'incinération des résidus
urbains, dont les principaux sont les émissions dans l'atmosphère et la
pollution, qui peut être engendrée par le dépôt des cendres volantes.
Le recours à des équipements appropriés et l'application de mesures de
sécurité peuvent contribuer à les réduire considérablement.

Le document decrit égalementsuccinctement les nouvelles techniques
d'utilisation énergétique des résidus urbains, qui en sont toutes encore
au stade expérimental.

La conclusion indique un certain nombre d'orientations et de recommendations, ainsi que quelques sujets de recherches et de développement à
approfondir.

ZUSAMMENFASSUNG

Das Referat liefert Daten über den in der Europäischen Gemeinschaft
anfallenden Hausmüll (95 Millionen Tonnen) und seine Zusammensetzung. Die
Konventionelle Verbrennung ist bisher immer noch die einzige praktikable
Methode zur Energierückgewinnung, doch werden bisher nur 13 % des Hausmüll
in dieser Weise verarbeitet. Es besteht jedoch ein allgemeiner Trend zur
grösseren Energiegewinnung aufgrund der steigenden Kosten der Energieträger
und der Knappheit an Gelände für Deponien. Einige Anlagen haben fast das
Stadium der Rentabilität erreicht.

Die Umweltprobeleme im Zusammenhang mit der Verbrennung von Hausmüll
werden erörtert, mobei das Hauptproblem die Schadstoffemission in die
Luft und die Umweltverschmutzung durch Flugasche sind. Diese Nachteile
können durch geeignete Ausrüstungen und sichere Betriebsverfahren
beträchtlich reduziert werden. Ferner werden die noch im Versuchsstadium
befindlichen neuen Techniken der Energiegewinnung aus Hausmüll beschrieben.

Als Ergebnisse werden einige Leitlinien und Empfehlungen sowie mögliche
kommende fuE-Bereiche skizziert.

Some 95 million tonnes of municipal waste were generated in the European Community in 1977, about three-quarters of which was composed of household waste. The rest was from offices and shops, industrial waste which can be classed with urban waste, market waste and street sweepings and large discarded objects.

Some of this waste, particularly paper and board, is recovered and recycled, either by selective collection or - more rarely - by mechanical sorting. This tonnage is equivalent to 24.2 million tonnes of high quality coal (TEC = 29 300 kJ/kg or 7 000 kcal/kg).

The total tonnage may rise to 100 million tonnes in 1980, but the trend is towards a levelling out of growth.

Another trend, since the last war, has been a marked growth in the calorific value of municipal waste. This is due to smaller tonnages of inert materials (combustion ash), largely because of reduced coal burning and because of an increase in discarded paper and board and, particularly since 1970 plastics.

This pattern, together with the rising costs of energy-producing raw materials, now makes it more attractive to recover energy from the incineration of municipal waste.

At present, all Member States face problems of disposing their municipal waste. Alternative disposal methods include :

- selective collection with recovery components such as paper, glass and plastics ;
- landfilling without crushing - still the most economical method. (2 to 9 EUA/tonne of waste) ;
- landfilling with crushing, costing 4 to 18 EUA/tonne ;
- conventional incineration; the main subject of this document ;
- composting for fertilizer, (11 to 27 EUA/tonne of waste) ;
- mechanical sorting : components may be recycled but the materials are soiled.

There is sometimes a wide variation in disposal costs. For example, controlled landfilling is dear in Denmark (14 - 18 EUA/t) and relatively cheap in the U.K.

Additional new disposal techniques now being tried out in pilot schemes or already on industrial scale include : high-temperature incinceration pyrolysis and gasification ; the manufacture of waste-derived fuel by a variety of methods including mechanical sorting, and biological treatment with methane production.

Since 1950 conventional incineration has become very advanced with an efficiency now reaching about 70 per cent in steam-raising. Further major improvements are unlikely. Grate furnaces are more suitable than rotary furnaces, which are also limited in throughput to 10 tonnes/hour.

Improvements to incinerators include increased automation, flue-gas cleaning, and increased energy generation In addition installations for breaking up large articles and separating slag and ferrous scraps have been developed and environmental protection measures have been improved.

Stoppage times caused by breakdowns are now much shorter but incinerator with energy recovery must be equipped with at least two furnaces or an auxiliary oil fired boiler.

The use made of the combustion heat depends essentially on local conditions. The steam may be used in large installations to generate electricity and in small installations to generate heat for drying sewage sludge for district heating or for industrial purposes.

Currently incineration is an expensive operation for use when landfill sites are not available, but the situation may change with increases in the cost of fuels. A secondary source of revenue is sometimes the sale of materials such as slag and ferrous scrap.

Conventional incineration also has the advantage of considerably reducing the initial volume of municipal waste down to 12 to 17 per cent and eliminating putrescible organic materials. Where high temperature incineration is used, as in the pyrofusion installation in Luxembourg, the reduction in volume is even greater (3 to 5 per cent) of the initial volume.

In 1977 about 22.2 million tonnes of municipal waste were incinerated in the Community, that is about 23 per cent of the total. More than 12.5 million tonnes were incinerated with energy recovery.

Population   Method

              1. landfill sites available ——→ Controlled landfilling

A   20 000   2. no landfill sites ——————→ Small incinerator without
                                                     energy recovery

B   20 000   1. landfill sites available ——→ Controlled landfilling

      to   2. market available ⟵————————→ Composting
                                          ——————→ Selective collection
    100 000   (mechanical sorting)
                                                  Incinerator with heat
                                                  recovery for heating or
                                                  industrial use

              3. no site, no market ——————————→ Incinerator without heat
                                                     recovery

C   100 000   1. landfill sites available ——→ Controlled landfilling
                                                  Composting
              2. no landfill site but ⟵————→ Incinerators with energy
                 market available   recovery :
                                                  hot water
                                                  steam
                                                  electricity
                                                  steam/electricity
                                                  Recovery of materials by
                                                  selective collection or
                                                  mechanical sorting

              3. no site, no market ——————————→ Incineration without
                                                     energy recovery.

Table II

Balance-sheet of the quantities and composition of municipal waste and the like arising in the EEC
(% by weight 1977)

Composition of domestic refuse	UK %	D %	F %	DK %	I %	IL %	B %	L %	NL %	EEC quant. (x10⁶ t)	% by weight
Paper and cardboard	30	27,5	35	35	30	33	30	25	23	22,07	29,9
Putrescible matter	21	16,0	20	15	35	28	43	56	48	18,07	24,6
Plastic	3	4	5	4	4	4	5	4,6	6	2,97	4,0
Textiles	3	3	4	2	3	3	1,5	1,5	1,8	2,24	3,0
Metals	9	6,5	5	4	3,5	4	4,5	3,6	2,9	4,39	6,0
Glass	9	9	8	8	5	8	10	5,2	13,0	6,06	8,3
Dust and fine debris	19)34	)34	10)13	)32	19,5)20	)20	)6	)4,1	)5,3	)17,80	)24,2
Other	9)	)	13)	)	20)	)	)	)	)	)	)
Total quantity Domestic of waste (x 10⁶ t)	19.0	19.0	12.0	1.23	14.5	1.2	3.0	0.13	3.5	73.56	100 %
Domestic and the like	25	25	(16)	2.53	16.3	(1.3)	(3.5)	(0.13)	5.3	(95.0)	
Quantity of waste incinerated/ year (x 10⁶ t)	2.3	5.44	3.2	4.9	5.8	–	0.54	0.1	1.74	22.23	
of which, with energy recovery	0.9	5.31	2.6	2.9	0.6	–	0.2	0.1	1.34	12.53	
mean calorific value kJ/kg	9400	7300	8400	8400	6700	9400	7300	7500	7500	7.500	
Kcal/kg	2250	1730	2000	2000	1600	2250	1750	1800	1800	1.800	

For energy recovery the situation was as follows in the Member countries
in 1977 :

	L	DK	NL	D	F	B	I	GB
per cent of total waste used for energy recovery	77.	47.	38.2	21.2	18.1	5.7	3.7	3.6
per cent of total waste incinerated	100.	86.	77.	97.5	59.	37.	10.3	39.

These figures show that energy recovery from the incineration of
municipal waste is still low in many countries. There are still too many
incinerators which do not recover energy, some of them large and energy
consuming.

The incineration of municipal waste makes only a small contribution to
energy savings in the Community. The energy value of municipal waste
incinerated with energy recovery in 1977 was about 22.6 million Gcal, which
corresponds to a saving of 3.3 million tonnes of coal (1.32 per cent of
coal consumption) or about 2.2 million tonnes of oil (0.42 per cent of
consumption of liquid hydrocarbons). The situation varies widely,
however, from one Member State to another.

Seasonal fluctuation in demand poses a problem when incinerators are
used for heating with, generally speaking, 60 per cent of the heat being
utilized.

Electricity generation is technically viable for continuous incineration
plant receiving over 100 t of waste per day. Yet in some countries the
sale price of electricity to the grid is often low, although when supplies
are interrupted, some independent plants have to pay a high price for
electricity supplied by the grid.

In France this problem has been largely solved by a ministerial decree
which lays down the terms of purchase by the EDF of electricity supplied
by independent producers. At Community level, appropriate legislation
would be desirable.

In 1977 municipal waste incineration residues may be estimated at some
7.5 million tonnes, broken down as follows :

    5 450 000 t of slag   ) clinker
      950 000 t of ferrous scrap)
    1 100 000 t of fly ash and processing sludge

Most of these residues, except some 300 000 tonnes of ferrous scrap and a very small amount of slag, is landfilled. Ferrous scrap after incineration is of mediocre quality, containing metallic impurities (Ti, Sn, Cr, Cu, etc) and residual slag. The presence of these impurities makes it difficult to use in blast furnaces, especially old plant. Prices vary with the fluctuation of demand : 0 to 24 EUA/t.

Slag after the removal of ferrous scrap is still very little used in the Community. It can be used for road foundation and experiments are in course on mixture with bitumen. Only a very slow start is being made in the use of incineration slag and road contractors are ill-informed about its use. The cost of slag may be estimated at about 4.2 EUA/t, including preparation costs.

Incineration fly ash in particular (from the boiler and gas cleaning system) and sludge from waste water treatment are not reused. They are usually, but mistakenly disposed of together with clinker on a landfill site. They are a hazard to the environment.

Greater use of incineration slag to replace natural materials would reduce the volume of waste, but the operation is viable only for local requirements. The incineration slag is also in competition with other types of slag or waste. The question of using ferrous scrap was studied in the Federal Republic of Germany. The iron-steel industry will take these scraps.

Total costs, i.e. annual capital and operating costs less any revenue, for the different methods of disposing of municipal waste fall in the following range :

	EUA/tonne of waste (1978)
controlled landfilling without crushing	2 - 9
controlled landfilling with crushing	4 - 18
composting	11 - 27
incineration without energy recovery	10 - 30
incineration with recovery of energy and possibly materials	7 - 30
materials recovery systems	(11 - 25)

As a result of the rising cost of energy, several incineration plants in the Community are at present near the stage of profitability.

For installations with energy recovery, total costs vary but generally
decline with the increase in the quantity of waste treated. They depend
on a series of factors such as thermal efficiency, receipts obtained,
optimum location and hourly capacity.

Some modern installations with energy recovery can now compete on economic
grounds with controlled landfilling after crushing. Profitability should
improve further with the probable increase in the cost of energy raw
materials.

In general the trend in several Member States is towards :
-     more incineration of some plants to include energy recovery ;
-     conversion of some plants to include energy recovery ;
-     replacement of polluting and less profitable old plants ;
-     additional capacity for some existing plants.

The United Kingdom is particularly interested in the manufacture of waste
derived fuel which can be used in certain industrial furnaces and perhaps
in some power stations. Ireland, which is not short of space, is more in
favour of landfilling. Italy has opted rather for a policy of materials
recovery. This approach is also followed, together with energy recovery,
in several other Member States.

Besides conventional incineration, new methods of treating municipal waste
with energy recovery are also being developed in most of the Member States.
In the Federal Republic of Germany a 50 per cent increase of energy
recovery from municipal waste is foreseen by 1990.

While most municipal waste incineration plants comply with national
environmental protection requirements, they nevertheless cause problems
of emissions nuisances such as noise and smells, and certain solid
residues which may be a potential source of pollution after being
landfilled. Compared with fossil fuels; the problems of emission into
the environment are more serious, except for SO_2 and NO_x.

The two main problems of emission into the environment from the incineration of municipal waste are air pollution and the pollution which may
be generated by the deposit of fly ash and processing sludge.

The principal substances emitted into the atmospher are :
-     dust composed mainly of various compounds and oxides. They average
      about 7 per cent of heavy metals in compound form (Zn, Pb, Cu, Cd,

etc) ;

harmful gases such as SO_2, NO_x, CO, HCL, H_2S, polycyclic hydro-
carbons.

Most incineration plants are equipped with flue gas purification systems
which efficiently remove dust but do not affect emissions of harmful gases.
Dust is removed from the gases by means of cyclone separators (effective
for particles of more than 5 microns) and electrostatic filters (sometimes fabric filters) with average efficiency of 98-99 per cent and most
effective for particles of 0.1 to 20 microns.

Dust emissions from the incineration of municipal waste in the Community
may be estimated at between 15 000 and 30 000 t per year. Although this
dust is only a small contribution factor in total air pollution in the
Community, it is rich in heavy metals like Zn, Pb, Cd and Cu (average
7 per cent of heavy metals in fly ash), some of which are partially
leachable.

As regards harmful gases, since 1970 there has been a considerable
increase in emissions of HCL and HF caused by the increase in the proportion of plastics in municipal waste. The emissions of other gases
(SO_2, NO_x, CO (CH) n) are for the most part below the limit values laid
down in several Member States. Emissions of HCL and also HF are of
particular concern in the Federal Republic of Germany. They may be very
much reduced by means of wet-method gas cleaning equipment (scrubber) or
by a dry method, injection of CaO in the gas circuit; this method is now
being tested.

Not much is yet known about emissions of organic micropollutants, including
dioxin compounds and polychlorobiphenyl (PCB). More research is needed on
their possible effects on human beings. Preliminary investigations done
in Italy indicate a very low annual emission level of dioxin compounds
(about 1 gr) from an incineration plant at Milan.

Generally incineration residues - clinker, fly ash and processing sludge -
are often mistakenly disposed of together at a landfill site where they may
represent a potential source of pollution. In fact, the behaviour of
clinker under atmospheric conditions is different from fly ash and
processing sludge. Apart from iron oxides, clinker contains up to 1.5 per
cent of heavy metals (Ti, Sn, Cr, Cu, etc). Experiments done in Denmark
and in the Grand Duchy of Luxembourg show that leaching rate by water is

low in comparison with the other residues and that most of the soluble
salts including some metallic compounds could be extracted after 2 days
treatment by water on ground material. Some ferrous scrap from incineration is reused in the steel industry for the manufacture of cast iron
and low quality steel but this reuse poses an environmental problem. On
the other hand, once the ferrous scrap has been removed, slag can be
reused for road building, but not enough use is made of it in the EEC.

Fly ash and processing sludge are a much greater hazard to the environment as certain harmful metallic elements, particularly Zn and Cd are
easily leachable by rain-water. They also contain soluble salts in
particular chlorides and sulphates in fairly high concentration. From
their differenices in behaviour separate landfill sites for clinker and
for fly ash and sludge are highly advisable.

In pyrolysis, the organic components are broken down into liquids (tars
and oils), gases (H_2, CH_4, and CO) and a carbon-rich residue. The
recovered gas is fractionated and cleaned of harmful substances (such as
dust, SO_2, H_2S, HCL) and may be either burned or stored. After separation,
the oil and tars may be used as fuel or further recycled. Metals and
glass can be recovered from the solid residue.

High-temperature incineration (Andco Torrax process)
High-temperature incineration is a process for treating municipal waste
which is already being used on an experimental scale in the Grand Duchy
of Luxembourg. Other plants exist at Frankfurt, FRG and at Grasse and
Creteil, France.

It combines pyrolysis, i.e. the thermal breakdown of the waste, with the
fusion of the solid residues obtained. The pyrolysis gases, which have a
calorific value of 4 000 to 6 000 kj/Nm^3 are burnt with the addition of
air in a combustion chamber. The hot gases, produced at temperatures of
1 000° to 1 300°C, contain little dust and are channelled into a heat
recovery boiler.

High-temperature incineration, which has a fairly high thermal efficiency
(73 to 74 per cent), sharply reduces the initial volume of the waste
(3 to 5 per cent), and no unburnt material is left in the cinders, but
there are still operating problems. The fly ash is also richer in metals
than the fly ash of conventional incinerators.

Waste derived fuel (WDF)

Waste derived fuel can be produced in a wide variety of different methods :
course shredding without or with magnetic separation - screening of the
shredded waste - mechanical sorting using screening and air classification.
This last leaves a light fraction (WDF) consisting mainly of paper, board
and plastics, a heavy fraction containing recoverable glass and metals
and a residue consisting mainly of fine dusts.

This waste derived fuel which may be pelletized will be more homogeneous
than crude municipal waste and will have a calorific value considerably
higher than the original input material (12 000 to 16 000 kj/kg). Its
calorific value may be further enhanced by enrichment which residual oil,
coal dusts and other additives.

There is likely to be a family of waste derived fuel with a range of
calorific value and with different handling and burning characteristics.
Their preparation costs are themselves relatively expensive and some of
the most sophisticated fuels especially those which have been pelletized
will be able to be stored for lengthy periods. The simpler fuels will be
best used to supplement coal in industrial and similar boilers and in
cement kilns. The more sophisticated fuels may eventually be usable in
power stations. The possible risks to high technology equipment means
that much careful proving work will be necessary.

Besides technological improvement, the health and ecological aspects of
using waste derived fuel require more study. Many member countries are
interested in producing waste derived fuels and a number of prototype
plants using different processes have been built.

Biological treatment

Experiments are in course in the Netherlands for the recovery of methane
gas produced by the anaerobic decomposition of municipal waste in sealed
landfill. It has been estimated that around 125 m^3 of gas (50 per cent
CH_4 - 50 per cent CO_2). The major proportion of methane will probably be
produced during the first 10 to 20 years after the sealing of the site.

Methane fermentation, particularly by anaerobic process would be of
particular interest for agricultural and livestock waste which arising
reaches some 900 million tonnes in the Community.

ENERGY FROM WASTE:

EUROPEAN INCINERATING TECHNOLOGY

Mogens Rasmussen

A/S Vølund

Denmark

SYNOPSIS

History of European incineration technology with energy recycling, leading to today's technology.

Total number of incinerator plants with energy recycling today amount to 347 furnaces with a capacity of 75 000 tonnes per day.

Type of energy recycling differs from country to country, but majority of plants are producing pressurized hot water, low or high pressure steam for heating purposes, electricity generation, or for industrial processes.

Refuse, an uncontrollable kind of material giving cause to problems with respect to corrosion and environmental considerations. The means for avoiding or minimizing those problems.

RESUME

Historique de la technologie européenne d'incinération avec récupération de l'énergie et situation de cette technique aujourd'hui.

Le total des usines d'incinération avec recyclage est aujourd'hui de 347 fourneaux, soit une capacité de 75 000 tonnes par jour.

Le type du recyclage produisant de l'énergie diffère d'un pays à l'autre, mais la majorité des usines produisent de l'eau chaude sous pression pour le chauffage, la production d'électricité ou les usages industriels.

L'experience montre qu'en Europe les installations sont sûres et largement disponsibles et on citera quelques exemples des disponibilités d'usines d'incineration modernes.

Les déchets, un matériau que l'on peut difficilement controller. Problèmes de corrosion. Protection de l'environnement. Moyens permettant d'éviter ces problèmes.

ZUSAMMENFASSUNG

Uberblick vom Beginn der Müllverbrennungs technologie mit Energierückgewinnung in Europa bis zum heutigen Stand der Technik.

Zur Zeit bestehen 347 Verbrennungsanlagen mit Energie-Recycling mit einer Gesamtkapazität von 75 000 Tonnen pro Tag.

Die Art der Energierückgewinnung ist von Land zu Land verschieden ; in den meisten Anlagen wird heisses Druckwasser, Nieder-oder Hochdruckdampf für Heisswecke, Elektrizität oder Prozesswärme erzeugt.

Die europäischen Anlagen haben sich hinischtlich Zuverlässigkeit und Verfüg barkeit bewährt, und in dem Referat werden einige Beispiele für die Verfügbarkeit moderner Müllverbrennungsanlagen gegeben.

Mit dem Müll muss eine Materie von unkontrollierbarer Zusammensetzung verarbeitet werden, wobei Probleme im Zusammenhang mit der Korrosion und dem Umweltschutz auftreten.

Mittel zur Vermeidung oder Beherrschung dieser Probleme ab.

Mass burning of refuse with energy recovery is a fully accepted way of disposing of residential, commercial and industrial refuse in Europe. The first known municipal incinerator was installed in Nottingham, England, in 1874 and was soon to be followed by many more. The first primitive incineration plants were of the batch fed, multiple hearth type, a principle which sets narrow limits to the maximum practical size of installations. Usually there was no recovery of energy, but some very early incinerators produced steam and/or electricity from the beginning of this century.

Scarcity of land and diseases such as the cholera epidemic in the second half of the last century caused a new hygienic conscience to develop, and one of the results of this was the organization of refuse collection and disposal in most European cities.

In the 1920's a technical development took place which would open the way for modern incineration design. The breakthrough was the construction of the refuse disposal plant at Gentofte, a surburb of Copenhagen, incorporating the world's first continuously operating refuse incinerator. The plant started operation in 1931 and was followed the following year by a similar plant in the Frederiksberg suburb of Copenhagen.

Both of these plants made use of the heat released by the incineration of refuse. At Gentofte electricity was generated and at Frederiksberg steam was supplied to the town's district heating scheme. Both plants, utilizing the Vølund system, were in operation for over 40 years.

The Copenhagen plants pioneered the way for much larger plants. The incorporation of a mechanical refuse transport on moving grates facilitated a closely controlled combustion, which in turn made it possible to achieve high standards for burn out and a high level of general hygiene.

The years after the second World War saw a massive investment on the European continent in incineration plants with heat utilization. For example in Denmark, approximately 65 per cent of all residential, commercial and industrial refuse and waste is disposed of by mass combustion with energy recovery.

Denmark is a country without any indiginous fuel resources and there is, consequently, a tradition for conserving energy and recovering waste heat. Energy recovery made even more sense because the expansion of refuse burning facilities coincided with tightening air pollution standards. Filters and electrostatic precipitators were becoming necessary, but the use of this equipment required a cooling of the combustion gases and it made good sense to achieve this cooling in boilers, thus converting the energy into high pressure hot water or steam.

Energy recovery from refuse was in many European countries assisted by the fact that the municipalities are responsible for disposal of refuse as well as for the supply of electricity, district heating, water and gas.

The following schedule gives an up-to-date survey of European incineration plants with energy sale in operation today, including Vølund installations. It also lists specifically the number of Vølund plants with energy recovery and sale built in Europe.

European Incineration Plants with Energy Sale

Country	Number of Plants	Number of Units	Sh tpd
Austria	2	5	1,450
Belgium	2	5	1,450
Denmark	31	48	8,690
Finland	2	4	380
France	22	42	12,600
Italy	15	30	5,700
Luxembourg	1	2	530
Netherlands	4	18	1,530
Norway	1	1	190
Spain	3	4	1,210
Sweden	15	31	5,040
Switzerland	19	43	6,750
UK	6	15	4,350
West Germany	40	99	35,000
Europe Total	163	347	84,770

Steam is the traditional heat medium in most European countries. It is not, however, the most suitable medium for long distance transfer of energy, for which high pressure hot water is more convenient.

Obviously, the choice of heat medium depends on how the energy can be best utilized. The following schedule reveals some of the possible utilizations of energy from refuse incineration and the heat mediums used :

HEAT MEDIUM   APPLICATION

Medium-pressure hot water
Flow temp. 80-120°C (175-248°F)   District heating for space heating,
                                              air conditioning and hot water
                                              services.

High-pressure hot water
Flow temp. 120-200°C (248-400°C)   District heating for industrial
                                              purposes and for space heating, air
                                              conditioning, hot water services and
                                              absorbtion cooling.

Low and medium-pressure steam
0-15 bar (0-225 psi)   District heating. Industrial
                                              process steam. Absorption cooling
                                              (ice production). Desalination
                                              Inplant steam-driven equipment.

High-pressure steam
15 bar - (225 psi -)   Electricity production in fully
                                              condensing turbines. Cogeneration
                                              in extraction turbines or back-
                                              pressure turbines (combined
                                              production of power with steam
                                              or hot water for uses as shown
                                              above).

Power generation has an inherent poor cycle efficiency and this is further reduced by the practical limitations to pressure and temperature which, I believe, everybody today accepts for refuse fired boilers. For the best energy efficiency as well as for optimum economy, the sale of energy for space heating purposes or industrial processes therefore is preferable. In Scandinavia the heat medium most widely used for this is high-pressure hot water.

The design of a modern incineration plant with energy recovery involves a combination of mechanical handling, combustion and boiler technologies. Today, with more than 50 years of experience in continuous operating refuse incineration plants with energy recovery, most problems have already been encountered and accounted for.

Five European manufacturers have among them built approximately 85 per cent of the energy recovering plants in Europe and it is safe to assume that all of these manufacturers are capable of building reliable plants with good availability. The design philosophy behind most plants unanimously favours rugged and reliable equipment for all the mechanical handling purposes as well as for auxiliary equipment.

Some difference of opinion has been seen in plant design, with high fuel-to-energy efficiency competes with high reliability. Vølund maintains a design incorporating a refractory lined incinerator with a separate boiler unit. Some other manufacturers prefer an integrated incinerator-boiler design, known as a water wall incinerator.

This plant design is generally based on experience gained from boilers for conventional solid fuels and does not put the same emphasis on the special thermal conditions applicable to the incineration of solid waste. Thus, we find that these systems tend to reflect primarily traditional, desirable boiler design requirements such as : high efficiency ; high pressure stability (the ability to withstand the required static pressure on the water/steam side with minimum use of material in boiler tube walls); good steam quality (no water droplets).

The most dangerous hazards to an incinerator reliability occur on the gas side of the furnace-boiler system, i.e. : fluctuating gas atmosphere; fouling; erosion; dew point corrosion; high temperature corrosion.

Dew point corrosion, in plants with heat utilization is rare in boilers or in auxiliary equipment. The exhaust gas temperature can easily be maintained well above the dew point temperature for acids in the flue gases.

High temperature corrosion, on the other hand, presents a serious threat and the following conditions are to be avoided :

-   The presence of local streaks of incompletely burned gases in the gas passages of the boiler.

-   Boiler wall temperatures (metal temperatures) exceeding 350-400°C (650-700°F).

-   The presence of a layer of fly ash or clinker in a melting phase on the boiler surface.

Recent investigations indicate that the most dangerous conditions exist when incompletely burned out gases get in contact with the boiler walls, thereby causing fluctuations between oxidizing and reducing atmospheres in the presence of high temperatures and corrosive gases. If these streaks of reduced atmosphere can be avoided then the metal temperatures in itself seems less important. The occurrance of melting temperatures in the fly ash and clinker layer, too, is often caused by this local combustion of unburned gases raising the temperature locally above the melting point.

It is, therefore, very important to avoid the streaks of reducing atmosphere in the boiler. This problem must be solved before the gas reaches the boiler rather than in the boiler itself.

Despite all efforts to mix the waste properly before it is fired, it remains a very heterogeneous fuel giving local streaks of unburned gases with high carbon monoxide content, and temperature fluctuations immediately above the grate, even for the presence of excess air. The conditions often are promoted by the very wide grate areas necessary in high capacity incinerators.

Gases only mix effectively when they are of the same temperature. There-fore, the combustion gases must be retained in the combustion zone long enough to ensure that they are completely burned out and properly mixed so that a homogeneous oxidizing atmosphere is created prior to entering the boiler.

The two-way gas system and the special after burning chamber allows the time, temperature, and turbulence necessary for complete combustion of the gases before they enter the boiler.

The fly ash particles consist of easily meltable clinker, which remains soft down to a temperature of approximately 600°C (1 100°F). Even after the surface of the fly ash particle is cooled below that temperature, the centre remains soft for some time, increasing the risk of particles sticking to the boiler tubes when they flatten on impact.

The degree of clinker slagging and sintering is often the decisive factor in determining when an incinerator must be taken out of operation for maintenance. Therefore, it is important that fly ash particles are burned up completely and effectively cooled down before entering the convection part of the boiler, where the boiler tubes are positioned. The first objective is achieved in the afterburning chamber. The second is met by designing the gas passages to allow sufficient time in the radiation zone of the boiler.

These objectives, are best achieved through a design incorporating furnace and boiler separate from each other. Compared with the integrated boiler design (water-wall-furnace) the separate furnace and boiler design sometimes require a marginally larger area and, in addition, the heat recovery, in theory at least, is of marginally lower efficiency.

Theoretically, the water wall incinerator furnace can be operated with less excess air than the refractory wall incinerator because no air cooling is required for the furnace walls. It is this theoretical reduction in excess air that has produced a marginally higher fuel-to-energy efficiency. However, existing water-wall-furnaces in operation in Europe have experienced severe erosion and corrosion problems in the water wall sections. This has largely been the result of the previously mentioned fluctuation between oxidized and reducing gas atmospheres in the furnace, but has also been influenced by the steam temperatures chosen for the boiler design.

The immediate effect of the problem of water wall corrosion is that plant operators and designers tend to increase substantially the amount of excess air, resulting in the elimination of the theoretically higher boiler efficiency. A more long term effect is that many manufacturers of water wall incinerators now have begun to install refractory linings inside the water walls in the furnaces. Furthermore, specifications for maximum steam outlet temperatures from the incinerators are now frequently being downgraded to 650-750°F following the many incidents of serious superheater corrosion experienced over the last ten years.

Naturally reliability and availability will vary from plant to plant depending upon design criteria, boiler temperature, general layout, success in choosing reliable auxiliary equipment and, not least, the care and skill of the operators.

In one case, the St. Quen incineration plant in Paris, we even achieve as much as 119 per cent of nameplate capacity on an annual basis.

Stand-by facility

The question of stand-by facility may be considered from the points of view of disposing of the refuse on a continuous basis, or of providing energy on an uninterrupted basis.

It is good design practice to split the refuse burning capacity between a number of units. For larger plants operating without a nearby "colleague" plant, three or four lines will provide flexibility to accommodate seasonal variations in the refuse quantity as well as allowing for scheduled maintenance. Such a design gives a very high degree of ability to handle the refuse under all circumstances.

Two lines will normally be adequate and even one line may be acceptable provided this line is not stretched to its full capacity. Short term variations in refuse quantity as well as loss of furnace capacity as a result of breakdowns can be accommodated by storing the refuse for a short period in the refuse pit, normally of two to seven days refuse capacity.

Each furnace line will normally be designed as an independent unit with its own boiler, precipitator and auxiliary equipment, the only common equipment being the refuse handling crane.

The need for stand-by facilities for reasons of continuous energy supply is determined by the type of energy produced, the customers connected to the plant and the organization of ownership of the plant.

By dividing the refuse burning capacity into several lines a certain safety factor is achieved, but this will, of course, be of no avail in the case of interruption of "fuel" supply as a result of a strike or adverse weather conditions.

For electricity producing plants the insignificant power output of an incinerator compared to a major power station is normally not of any consequence for the requirements of a large electricity grid. Likewise, for larger district heating schemes the refuse burning plant will normally represent a base load.

If the refuse burning facility is the sole supplier of energy for one or more industrial customers, a stand-by facility will be necessary, whose cost will normally be comparatively modest.

The rotary kiln

The rotary kiln has been a significant part of the Vølund furnace design since the first plant was intalled in Gentofte in the late 20's. The rotary kiln serves as the last part of the grate in the larger plants, but is not normally economically justified for smaller plants, say below 5-6 tons per hour capacity. The rotary kiln provides a very high degree of flexibility and, thus, allows the plant to handle refuse of widely varying compositions and heat values. It produces a residue in which heavy metals are bound in non-soluable compounds.

Environmental considerations

The flue gases from all modern incinerators are cleaned effectively through electrostatic precipitators, and therefore contain very small amounts of dust. A survey over more than five years carried out in Copenhagen in the years before and after the start of operation of the two big incinerators in the east and west of Copenhagen has not been able to trace any influence at all by these two facilities which incinerate 600,000 tonnes waste annually.

Nobody denies that the burning of PVC creates hydrogen chloride. However, assuming that the heat amounts produced by waste incineration should have been produced by fuel oil, the amount of acids in the air would have been 8 to 10 times higher.

Economics

The experience of the first eight years of operation has shown major economical achievements. The income from sale of heat and residues would cover more than the direct cost of operation, if the total heat production was sold. Today, only 2/3 of the production is sold, whereas 1/3 is cooled off.

In view of the very high oil prices of today, these returns have improved considerably and the total savings in oil purchase to the individual oil-fired installations would amount to 40 million Danish kroner. Should an equal facility have to be built today, the necessary investment would amount to :

Mechanical equipment	110 million D. Kr.
Building	100 million D. Kr.
District heating facility	15 million D. Kr.
Total	225 million D. Kr.

The annual direct operation cost amount to approximately 20 million D.Kr., leaving 20 million D.Kr. to pay off the investment.

Calculating with a treatment fee of 25 D.Kr. per ton waste, and depending on the rate of interest, the pay-back time of the investment will vary from

	14 years	(8 per cent)
to	25 years	(16 per cent)

If oil prices increase 30 per cent during the next two years, the pay-back period will be reduced to eight and 17 years respectively.

THE RECOVERY OF ENERGY FROM WASTE:

RESEARCH IN ITALY

Prof Giancarlo Chiesa

National Research Council

Italy

SYNOPSIS

In 1976 the National Research Council introduced the Final Energy Programme in response to the major developments in the energy supply situation.

One of the sub-projects is concerned with the recovery of energy from solid urban refuse and is aimed at collecting data and carrying out experiments on the recovery of energy from waste of interest to technical experts and administrators in this sector.

A study has been initiated and is still in progress on the quantity and quality of solid urban refuse and industrial waste in order to provide a detailed and up-to-date picture of the present situation.

Funds have also been provided for experimental plants to develop a number of recovery processes, i.e. : anaerobic digestion, pyrolysis and co-combustion at medium and large scale industrial plants.

It was found that one of the main difficulties to overcome with inciner-ation was that of atmospheric pollution, especially that caused by organochlorinated micropollutants (PCDD, PCDF, PCBs).

A survey has been undertaken to identify the presence of these compounds in the effluent from certain plants in Italy; concurrently a study has begun at a pilot plant to investigate the combustion kinetics of micro-pollutants and the mechanisms by which they are formed.

Noteworthy among other experiments carried out are schemes for the separate collection of paper and glass in five Italian towns (a total of some 200 000 people are involved) and field trials of compost in open fields and under glass which are now in their third year.

RESUME

_En 1976, à la suite des événements bien connus concernant les disponibili-
tés en énergie, le "Consiglio Nazionale delle Richerche" (Conseil national
de la recherche) a élaboré le projet finalisé "Energétique"._

_Un des sous-projets mis en oeuvre dans ce cadre porte sur l'utilisation
énergétique des déchets solides urbains et a pour objet de rassembler des
informations et de procéder à des expériences utiles aux techniciens et
aux administrateurs de ce secteur en vue de la récupération d'énergie à
partir des déchets._

_Une enquête sur la quantité et la qualité des déchets solides urbains et
des déchets industriels a été entreprise afin d'avoir une image détaillée
et à jour de la situation et elle est toujours en cours._

_En outre, des installations expérimentales ont été financées en vue de
mettre au point certains processus de récupération : digestion anaérobie,
pyrolyse et combustion auprès d'utilisateurs industriels de grande et
moyenne importance._

_En ce qui concerne l'incinération, on a constaté qu'une des principales
difficultés à résoudre est la pollution atmosphérique, en particulier
la pollution causée par des micropolluants organochlorés (PCDD, PCDF, PCB)._

_Une enquête a été entreprise en vue de constater la présence de ces compo-
sés dans les effluents de certaines installations en Italie, et, parallè-
lement, l'étude de la cinétique de combustion et des mécanismes de formation des micropolluants a débuté dans une installation pilote._

_Parmi d'autres expériences qui ont été effectuées, il y a lieu de rappeler
la collecte sélective de vieux papiers et de verre effectuée dans cinq
villes (près de 200.000 habitants étant concernés au total par cette
action), et le contrôle agronomique du compost en plein air et sous serre
qui en est à la troisième campagne agricole._

ZUSAMMENFASSUNG

Der "Consiglio Nazionale delle Richerche" hat 1976 infolge der
Entwicklung auf dem Energiemarkt das endgültige Energieprogramm erstellt.
Eines der einschlägigen Untervorhaben befasst sich mit der Verwertung
der Energie aus festem Siedlungsmüll; im Rahmen dieses Vorhabens sollen
Informationen verarbeitet und Versuche durchgeführt werden, die für die
Techniker und Verwaltungsfachleute dieses Sektors bei der Gewinnung von
Energie aus Abfall zweckdienlich sind.

Es wurde eine Untersuchung über die Menge und die Qualität von festem
Siedlungsmüll und Industriemüll durchgeführt; diese Untersuchung läuft
noch immer; anhand dieser Arbeiten will man sich ein genaues Bild von
der derzeitigen Lage verschaffen.

Ferner wurden Versuchsanlagen für die Entwicklung bestimmter
Rückgewinnungsverfahren finanziert: anaerobe Zersetzung, Pyrolyse und
kombinierte Verbrennung bei mittleren und grossen Industrieunternehmen.

Bei der Veraschung ist eine der Hauptschwierigkeiten die Beseitigung der
Luftverschmutzung, insbesondere der Verschmutzung infolge von Organochlor
Mikroschadstoffen (PCDD, PCDF, PCB).

Es wurde eine Untersuchung durchgeführt, mit der das Vorhandensein
dieser Verbindungen in den Abflüssen einiger Anlagen in Italien und
gleichzeitung bei einer Pilotanlage aufgezeigt werden soll; ferner hat
man Untersuchungen über die Verbrennungskinetik und die Mechanismen der
Bildung von Mikroschadstoffen aufgenommen.

Unter den durchgeführten Versuchen sind das gesonderte Einsammeln von
Papier und Glas in 5 Städten (es waren insgesamt etwa 200 000 Einwohner
an dem Vorhaben beteiligt) und die agronomische Überprüfung von Kompost
auf dem Felde und im Treibhaus, die nunmehr in das dritte Landwirtschafts-
jahr geht, zu nennen.

The problem of the disposal of domestic refuse is not nowadays just as a matter of environmental health, but also must be concerned with the recovery and reuse of some of its components.

Hence the National Research Council (CNR) in Italy has included an Energy Project and under that Project, has provided funds for numerous research reventures (to a total value of approximately 5 500 million lire over the five year period 1976 - 1980 for reclaiming energy from solid waste).

From the investigation conducted from 1977 on, it appears that the annual tonnage of urban refuse collected in Italy is approximately 14 million tonnes, corresponding to specific production of approximately 0.7 kg per person per day.

Of this, approximately 80 per cent is disposed of by tipping, some uncontrolled, about 15 per cent is incinerated and the residual 5 per cent is put through composting and recycling plant.

In italy three incineration plants, two in Milan and one in Genoa, are producing electricity. Total power generation is 120 million kWh/year.

Of special interest are the two recycling plants in Rome with potential capacity of 1 800 tonnes/day, and the 250 tonnes/day recycling plant in Perugia. In these three plants, paper, plastic, glass and ferrous metals are recovered and the organic fraction is recovered for use as a compost or feeding stuff.

On the assumption that 50 per cent of the tonnage of refuse produced could be submitted to recovery processes, Italy could count on an internal energy source, in terms of heating power, that would satisfy approximately 1 per cent of its national needs. Simple combustion does not achieve the maximum level of recovery of non-renewable resources. Even more attractive is the recycling of materials with high energy content. Hence research has been directed towards practical application of existing knowledge.

One factor that has emerged clearly has been the demand for reliable, consistent information on the physical properties and composition of domestic refuse and salvage techniques.

It is clear that recovery cannot be carried out free of charge. A technical and economic assessment must be made, based on experimentally tested data, to establish the viable level of efforts to recover waste, allowing for the fact that those efforts considerably complicate collection methods or may call for over-sophisticated disposal techniques.

The research programme has been broken down into the following main themes :

- investigation of quality and quantity of solid urban refuse and industrial wastes, and the existing methods of disposal combined with salvage ;
- investigation of chlorinated organic micro-polluting substances in the emission from the refuse incineration plants ;
- trials on the graded collection of paper and glass, run by the municipal authorities in certain medium sized towns ;
- combined combustion of pre-treated refuse and conventional fuels ;
- compost production and its testing for agricultural purposes ;
- anaerobic digestion of the organic fraction, with bio-gas production ;
- pryolysis of urban refuse and food industry waste.

A mobile laboratory was used for research on the quality of solid urban refuse. Visiting 30 locations, a representative sample of the country as a whole, it determined the composition of the local refuse and submitted it to a chemical analysis.

To this end, analyses were conducted for three to five days, at least 2.5 tonnes of refuse being screened per day. These measurements were repeated annually at different times of the year to determine the seasonal variations.

The data compiled up to this point are for 1977, 1978 and 1979. Summary data are given in tables 1 and 2.

Research on the quantity of refuse collected in Italy was conducted by means of a questionnaire sent to just under 2 000 local authorities in all towns with a population of over 5 000 and 100 local authorities for sample towns with a population of under 5 000.

Table 3 sets out the specific production of urban waste, the local authority areas having been graded according to population size.

Table 1 - Chemical and physical analysis of solid urban refuse

	Water %	Combustible Materials %	Incombustible Materials %	Net Calorific Value KCal/Kg.
Average weight per zone				
Northwest	40.3	36.3	23.4	1511
Northeast	41.3	34.9	24.2	1414
Centre	42.9	32.7	24.4	1334
South	48.9	29.0	22.1	1021
Islands	50.3	30.1	19.6	1128
Average weights per size of population				
20.000	45.5	30.7	24.0	1182
20.000 50.000	44.9	33.4	21.7	1306
50.000 100.000	43.1	33.9	23.0	1341
100.000 200.000	41.0	36.2	22.8	1472
200.000 500.000	40.7	35.6	23.8	1574
500.000	42.9	36.2	20.9	1450
Average weights based on main activity of area				
Agricultural/commercial	44.5	32.9	22.5	1269
Tourist/commercial	44.4	34.1	21.5	1333
Industrial/commercial/ tourist	42.0	35.0	23.0	1410
Industrial/commercial	41.7	36.5	21.8	1496
National average weights	44.2	32.9	23.0	1295

TABLE 2 - analysis - comparative average data, expressed as percen-
          tage in weight

	Sub-screen size	Cellulose	Plastic	Metals	Inerts	Organic Matter
	%	%	%	%	%	%
Average weights per zone						
Northwest	17.7	25.6	8.5	3.3	7.6	37.4
Northeast	16.2	23.1	7.3	3.4	9.6	40.4
Centre	18.0	23.3	6.4	3.0	8.6	40.7
South	21.1	18.5	6.6	2.5	5.0	46.3
Islands	13.3	20.4	8.3	2.2	7.9	47.9
Average weights per size of population						
20.000	18.5	20.9	6.6	2.8	7.3	43.9
20.000 50.000	18.7	21.9	7.8	3.0	5.9	42.7
50.000 100.000	17.7	22.5	7.9	2.9	6.5	42.5
100.000 200.000	17.8	24.0	7.3	3.5	7.2	40.2
200.000 500.000	20.3	23.6	7.7	3.5	7.6	37.3
500.000	16.9	25.7	7.6	2.9	7.8	39.1
Average weights based on main activity of area						
Agricultural/commercial	20.1	21.0	6.4	2.9	6.1	43.6
Tourist/commercial	17.4	23.1	7.3	3.1	6.9	42.3
Industrial/commercial/ tourist	17.8	24.8	6.3	3.4	10.1	37.5
Industrial/commercial	17.8	25.1	8.8	2.9	6.8	38.6
National average weights	18.3	22.3	7.2	3.0	7.1	42.1

Table III

Table of specific daily refuse production (Grams/ab/day)

Replies within the 250 – 1000 range or tourist resorts

Population range	Northwest	Northeast	Centre	South	Islands	Italy
Under 10000	612.1	623.9	654.6	672.8	521.5	624.2
10 – 30000	670.4	758.9	783.5	664.9	752.7	716.9
30 – 50000	597.7	618.5	657.9	659.1	723.1	639.5
50 – 100000	641.5	670.9	678.3	723.2	656.8	678.7
100 – 200000	660.6	657.3	732.0	555.9	602.7	664.0
200 – 400000	573.4	727.1	0.0	563.6	545.1	642.9
Over 400000	648.0	747.5	654.5	962.5	610.6	693.6
TOTAL	640.4	692.7	686.6	729.7	638.6	676.7

Parallel to the research on solid urban refuse, research is being
conducted to determine the quality and quantity of industrial waste
produced in Italy, with a breakdown according to economic sector. Data
are already available on the food and the engineering industries.

Tests have also been conducted on the operation of disposal + salvage
plants to determine their yield and running costs. In the case of
incineration plant, the investigation was broadened to determine the
polluting substances present in solid residues (slag and dust) and in the
liquid and gaseous effluent, including heavy metals and chlorinated
organic compounds.

For more thorough research on the mechanisms whereby chlorinated organic
micro-pollutants are formed, a pilot-scale incineration plant is being
built, equipped with appropriate instruments and with a post-combustion
chamber that can be used to determine the optimum operating conditions at
which this type of pollution is eliminated or kept as low as possible.

Graded collection trial

To find out whetherit would be technically and financially feasible to
collect sorted waste as a permanent refuse collection method administered
by the public authority, two types of trial were funded. They were
carried out in medium sized towns, each with a population of about
50 000 and consisted of :
- the collection of paper and glass by leaving bags with households ;
- the collection of paper by providing large crates (approximate
  volume 1.5 m^3) in the streets.

The most satisfactory results were obtained by the second method of
collection : 16 kg of paper was collected per head of population per
year.

The higher cost of collection was almost completely offset by the
proceeds derived from the sale of paper for pulping. The experiments
conducted in the various towns lasted for at least one year.

Two research products are being conducted on the use of solid urban
refuse as an additional fuel in large capacity power stations and in
special combustion chambers (tunnel kilns for bricks and furnaces for
cement).

The experiment on joint combustion in large capacity power stations was launched in the ENEL (national electricity company) 250 MWe power station at S. Barbara (Arezzo), the main fuel being lignite. Short tests show that up to 10 per cent solid domestic waste can be used. However first the ferrous metals have to be removed and then the product is subjected to grinding in the mills used for the lignite. A pre-treatment line is being set up (for removal of ferrous metals, removal of glass and grinding) at the power station for long term trials. If the results are satisfactory, including the findings on atmospheric pollution, the experiment will be extended to include coal-burning power stations.

Combustion tests are also being carried out on tunnel kilns for bricks, the ground refuse being injected into the brick slurry itself. Combustion tests are also being carried out on revolving furnaces for cement; the findings will be available late in 1980.

Compost production and agricultural trials

In proposing this research theme, the aim was not so much to develop composting technology but rather to assess the agricultural effects of the compost when used for different types of arable land and for different crops.

During the two farming years of 1977-78 and 1978-79, trials were conducted in open fields (over an area of approximately 30 000 m^2) and under glass, and it has been possible to compare the results with those achieved by the use of chemical fertilizers, animal manure and compost.

It appears that the use of compost derived from the decomposition and stablilization of the organic fraction of urban refuse helps to restore the organic substances removed by the crops to the soil. This is a very pressing need in Italy, where the organic content of agricultural soils is only 0.5 to 3 per cent, compared with the average European content of 2.5 to 6 per cent.

To conclude this brief review of research on the use of solid refuse financed by the National Research Council, it should be pointed out that construction work is taking place on a pilot plant for anaerobic digestion of the organic fraction of domestic refuse, and another pyrolysis plant. Experiments are also under way to assess the production of biological gas from the discharge of solid refuse and the viability of using that gas.

QUESTIONS & COMMENTS

Mr Rasmussen made the point that the capital cost of special boilers needed for burning RFD brought the total cost of the manufacture of RFD and its combustion almost to the level of straight refuse incineration.

Mr Smout said that RFD and WDF should be thought of as a family of fuelsThey could be simply waste that had been shredded and had the metals removed. Or they would be be fuels that had gone through a number of phases. There was need for research into RFD.

Mr R.G.Loram,UK, mentioned that large appartment blocks on the Continent were using RFD as fuel in their central heating plants.

Mr Jackson, of Warren Springs Laboratories, said that trials of burning RDF indicated no problems but the trials were so far only short term.

THE POLITICS OF WASTE PAPER
IN THE EUROPEAN COMMUNITY

Léon Klein

Environment and Consumer Protection Service

Commission of the European Communities

SYNOPSIS

*At present waste paper represents 40-50 % by volume and 15-20 % by weight
of urban waste, or 15-17 million t/yr; this causes a problem with the
accumulation of waste, pollution and disposal costs.*

*Less pollution occurs when waste paper is used in the manufacture of paper
products than results from the manufacture of pulp and of paper products
from it.*

*As estimated six-fold energy saving is made if paper products are manufact
ured from waste rather than pulp.*

*The Community today consumes 30 million tonnes of paper and board a year
of which only 14 million tonnes are produced from fibres originating within
the Community (5 million tonnes of wood fibre, 9-10 million tonnes of fibres
recycled from waste paper). Imports of paper, board and paper fibres
therefore amount to 16 million t/yr - a 6 000 million EUA deflict, second
only to that for petroleum products.*

*It is therefore highly desirable that more use should be made of waste
paper in order to reduce our dependence on outside sources in terms of
resources and currency.*

*Since 1974 the question of waste paper has been complic.ited by the fact
that the economic crisis has caused a considerable fall' in the price of
pulp and that, as a result, the price of waste paper is no longer attractive
this applies to many quality grades, especially the higher ones. It should
be noted, however, that since the fourth quarter of 1978 the price of pulp
has been rising.*

Recovery of waste paper has, of course, been affected by the crisis: it is now in a very difficult situation, as are many firms in the paper industry. There has been some improvement since 1979, however.

Increased utilization is possible, since the waste paper utilization rate in the Community paper industry is 40 % on average but could be raised to as much as 60 %. Admittedly, the Community average of 40 % is based on situations which differ widely from one Member State to the next; some have already attained a rate of approximately 50 %.

It is worth bearing in mind, however, that an increase in the average utilization rate from 43 % to 52 % would make it possible to use 12 million t/yr of waste paper instead of the present 9 million. Applied to annual production figure of 30 million tonnes of paper and board, by 1995 this rate would mean that about 15 million tonnes of waste paper would be used per year.

Measures

Action should therefore affect both the supply and demand for waste paper. In accordance with the guidelines adopted by the Waste Management Committee the following measures, for example, could be taken :

A. Demand:

    1) The use of recycled paper by public bodies should be promoted.

    2) Treatment techniques should be developed to enable paper prod·· ucts which contain a certain promortion of recycled fibres to compete, on both quality and price, with paper manufactured from virgin pulp (de-inking, decontamination).

    3) Specifications for the various paper products should be drawn up to enable the quality of manufactured papers to be better suited to their use and to avoid the presence of contaminated substances which would preclude or impede recycling operations.

B. Supply:

    1) The Waste paper recovery operator should be paid by the public authorities for collecting 'waste" (under certain conditions and within certain limits).

    2) Long-term supply contracts between recovery operators and manu- facturers should be encouraged.

    3) Investment which encourages the development of waste paper should be subsidized.

RESUME

_Les vieux papiers représentent actuellement environ 40 à 50% du volume et
15 à 20% du tonnage des déchets urbains, soit environ 15 à 17 millions de
tpa, d'où un problème d'entassement des déchets, de pollution et de coût
d'élimination._

_Par ailleurs, la pollution causée par la fabrication de produits papetiers
à partir de vieux papiers est moins grande que celle résultant du cycle
de fabrication des pâtes à papier et des produits papetiers à partir de
cette pâte._

_Enfin, on estime que l'économie d'énergie est dans un rapport de 1 à 6,
suivant que les produits papetiers sont fabriqués à partir du vieux papier
ou des pâtes à papier._

_La Communauté consomme actuellement 30 millions de tonnes de papiers et
cartons par an, 14 millions de tonnes seulement sont produites à partir
de fibres d'origine communautaire (5 millions de tonnes de fibres de bois,
9-10 millions de tonnes de fibres recyclées à partir des vieux papiers).
Les importations de papiers et cartons ou fibres papetières représentent
par conséquent 16 millions de tpa ou 6 milliards d'UCE (ce déficit est le
2ème en valeur après celui des produits pétroliers). Il est donc également très souhaitable d'augmenter l'utilisation des vieux papiers pour
réduire notre dépendance vis-à-vis de l'extérieur tant en devises qu'en
ressources._

_La question des vieux papiers est compliquée depuis 1974 par le fait que
la crise économique a fait baisser considérablement les prix de la pâte
à papier et que, par conséquent, pour bon nombre de ses qualités, surtout
supérieures, le vieux papier ne présente plus d'attrait sur le plan des
prix. Il est à noter cependant que depuis le 4ème trimestre 1978 les prix
des pâtes sont en hausse._

_La crise s'est évidemment répercutée sur la récupération de vieux papiers
qui est devenue très difficile en même temps que celle de nombreuses
industries papetières. Cette situation s'est améliorée depuis 1979._

_Une augmentation de l'utilisation : elle est possible puisque le taux
d'utilisation du vieux papier dans l'industrie papetière communautaire
est en moyenne de 40% et que l'on peut estimer à 60% l'augmentation_

*possible de ce taux. Il est vrai que la moyenne communautaire de 40%
d'utilisation recouvre des situations très inégales d'un Etat membre à
l'autre, certains Etats membres se situant déjà à 50% environ.*

*On peut noter cependant qu'une augmentation moyenne du taux d'utilisation
de 43% à 52% permettrait d'utiliser 12 Mio de tpa de V/P (contre 9 Mio
actuellement). Appliqué à une production de 30 Mio de P/C, en 1995 ce
taux ferait passer les V/P utilisés à ± Mio de tpa.*

<u>*Les Moyens d'Action*</u>

*Les efforts devraient donc porter à la fois sur la demande et sur l'offre
de vieux papiers. Conformément aux orientations prises par le Comité de
Gestion des Déchets, ces mesures pourraient être, par exemple :*

A. En ce qui concerne la demande :

    *1. Promotion de l'utilisation du papier recyclé par les services
publics.*

    *2. Développement de techniques de valorisation permettant aux produits
papetiers contenant une certaine proportion de fibres recyclées
d'être concurrentiels, du point de vue qualité et prix, avec les
papiers fabriqués à partir de pâtes vierges (désencrage, décontami-
nation).*

    *3. Elaboration de spécifications pour les différents produits papetiers
permettant d'une part d'ajuster davantage la qualité des papiers
fabriqués à l'usage que l'on en fait et d'autre part d'éviter la
présence de substances contaminées interdisant ou gênant les
opérations de recyclage.*

B. En ce qui concerne l'offre :

    *1. Rémunération par les autorités publiques, à certaines conditions
et dans certaines limites, des services rendus par le récupérateur*

    de vieux papiers pour le travail de collecte de "déchets" effectué.

    *2. Promotion de contrats d'approvisionnement à long terme entre
récupérateurs et fabricants.*

    *3. Subventions aux investissements favorisant le développement du
vieux papier.*

ZUSAMMENFASSUNG

Der Anteil des Altpapiers am Stadtmüll macht zur Zeit etwa 40 50 % vom Volumen und 15-20 % vom Gewicht aus, das sind rund 15-17 Millionen Tonnen im Jahr. Dadurch entstehen Probleme wegen der Anhäufung, der Umweltbelastung und der Kosten für die Beseitigung. Die durch die Herstellung von Papiererzeugnissen verursachte Umweltbelastung ist beim Einsatz von Altpapier weniger gross als beim Einsatz von Halbstoff.

Auch benötigt man bei der Herstellung von Papiererzeugnissen aus Altpapier schätzungsweise nur ein Sechstel für die Herstellung aus Halbstoff erforderlichen Energiemenge.

Die Gemeinschaft verbraucht derzeit 30 Millionen Tonnen Papier une Pappe jährlich; nur 14 Millionen Tonnen werden aus Fasern hergestellt, die aus den Gemeinschaftslandern stammen (5 Millionen Tonnen Holzfasern, 9-10 Millionen Tonnen Fasern aus der Altpapierrückgewinnung). Die Einfuhren von Papier und Pappeerzeugnissen oder von Papierfasern betragen somit 16 Millionen Jahrestonnen oder 6 Milliarden ERE (dies ist wertmässig der zweitgrösste Rohstoffbedarf nach den Erdölerzeugnissen). Es ist daher sehr wünschenswert, mehr Altpapier zu verwerden, um sowohl unsere Devisen- wie auch Rohstoffabhängigkeit von Aussen zu verringern.

Die Lage auf dem Altpapiersektor ist seit 1974 schwieriger geworden, weil infolge der Wirtschaftskrise die Preise für Papierhalbstoff erheblich gefallen sind und deshalb eine ganze Reihe von Altpapierqualitäten, insbesondere die besseren, vom Preis her nicht mehr interessant sind. Allerdings ziehen die Preise für Papierzellstoff seit dem 4 Quartal 1978 an.

Die Krise hat sich natürlich auf das Sammeln von Altpapier ausgewirkt, wo die Lage ebenso wie bei vielen Papierherstellern sehr schwierig geworden ist. Hier ist seit 1979 eine Besserung eingetreten.

Eine Steigerung des Einsatzes: Dies ist möglich, da die durchschnittliche Einsatzrate von Altpapier in der Papierindustrie der Gemeinschaft 40 % beträgt und eine Steigerung auf 60 % unterstellt werden darf. Allerdings spiegeln sich in dem Gemeinschaftsdurchschnitt von 40 % recht unterschiedliche Verhältnisse in den Mitgliedstaaten wider, wobei einige Mitgliedstaaten bereits bei rung 50 % angelangt sind.

Bei einer durchschnittlichen Steigerung der Einsatzrate von 43 % auf 52 % kämen immerhin 12 Millionen Jahrestonnen Altpapier zum Einsatz (gegenüber 9 Mio zur Zeit). Bezieht man diese Einsatzrate auf eine Produktion von 30 Mio Altpapier, so würde 19995 die Verwendung von Altpapier rung 15 Mio Jahrestonnen betragen.

Die Aktionsmittel

Die Bemühungen müssen sich also gleicherweise auf die Nachfrage- wie auf die Angebotsseite des Altpapiers richten. Nach den Leitlinien des Ausschusses für Abfallbewirtschaftung könnte es sich beispielsweise um nachstehende Massnahmen handeln:

A. Nachfrageseite:

    1) Förderung der Verwerdung von ERzeugnissen aus Altpapier in der Verwaltung.

    2) Weiterentwicklung von Verwertungstechniken, um Papiererzeugnisse mit einem bestimmten Gehalt an Altpapier von der Quailtät und vom Preis her gegenüber Papiererzeugnissen aus reiner Rohmasse wettbewerbsfähig zu machen.

    3) Ausarbeitung von Spezifikationen für die verschiedenen Papier-erzeugnisse mit dem Ziel, die Qualitaten besser dem Verwendun gszweck anzupassen und kontaminierte Substanzen auszuschalten, die eine spätere rückführung unmöglich machen oder behindern würden.

B. Angebotsseite:

    1) Unter bestimmten Bedingungen und in gewissen Grenzen Zahlung eines Zuschusses an den Einsammler von Altpapier durch die öffentliche Hand;

    2) Förderung langfristiger Lieferverträge zwischen Sammlern und Herstellern;

    3) Investitionsbeihilfen für Entwicklungsarbeiten auf dem Alpapiersektor.

Bei der KEG läuft eine Untersuchung über das Problem der Kontaminations-faktoren und der Spezifikationen für Papiererzeugnisse. Die Ergebnisse werden Ende 1980 vorliegen.

Recycling waste paper and board was the first priority of the Commission's Waste Management Committee set up in October 1977 as part of an "active policy to combat waste" approved by the EEC Council in May of that year.

The emphasis on paper and board should be seen against a background of member states consuming 30 million tonnes of paper and board per year, of which only 14 million are produced from fibres originating within the Community. Of the locally produced fibres, five million come from the native wood, and 9 to 10 million from recycled waste paper.

The imported material, 16 million tonnes per year, represents a financial deficit of 6 000 million EUA, making it second only to oil as an import of commodities.

It is therefore highly desirable that more use be made of waste paper in order to reduce EEC countries' dependence on outside resources. It is even more desirable because supplies from non-member countries may well become unreliable in the medium to long term.

The most optimistic forecasts suggest that the production of pulp in the Scandinavian countries, currently the Community's main suppliers, will increase by 35 per cent by the end of the century, that is by approximately six million tonnes. However this extra quantity is likely to be supplied in the form of paper rather than pulp.

South America as a potential supplier is dependent on the development of its internal consumption, but is not likely to be able to deliver more than one million tonnes per year before 1990. Africa will make no decisive contribution before 2000 through lack of investment.

Surpluses, if any, from Asian countries are likely to be directed to Japan. This leaves only North America and in particular Canada as the main suppliers between now and 2000.

These supplies are of marginal importance to North America compared with the domestic market, accounting as they do for barely 10 per cent of the production. But they do represent a considerable proportion of the Community's imports, and this dominance is likely to grow. Prices will at the same time be determined largely by internal demand, leading to a situation of potential instability.

The Community's dependence on external factors and the uncertainty of supplies makes it all the more urgent to develop a resource policy as soon as possible.

This policy should be conducted in step with research into alternative uses for waste paper, although the manufacture of paper and board must apparently remain the principal outlet for waste paper.

Improved recovery of waste paper from the 90 million tonnes of municipal waste generated in the Community each year could have a marked effect on the total amount of paper recovered as a proportion of the total consumed. In fact the recovery could rise from the present level of 32 per cent to as much as 55 per cent.

A major argument in favour of increasing the recovery of waste paper is that less pollution occurs when waste paper is used in the manufacture of paper products than when using virgin fibre. Also it saves energy, by a factor estimated to be as high as six. Finally, the development of waste paper recycling technology increases employment, in the recovery sector.

What then are the prospects for developing the consumption of waste paper, and the obstacle to be surmounted ?

The first fact to remember is that the Community's current need for fibrous paper material can be broken down into : paper and board - 8 million tonnes ; pulp - 7 million tonnes ; and wood - 1 million tonnes.

This 16 million tonne total will grow to 40 million tonnes by 1995 if the mean annual growth of consumption is at the rate of 1.6 per cent per year. A policy for Community   production which is aimed to increase the production of pulp in the Community and/or a policy for Community waste paper which aimed to reduce imports would make it possible to remove the wood deficit, and to reduce the pulp deficit, and cut the paper and board deficits by 20 per cent.

If these objectives, which seem realistic, are achieved, imports could be reduced by some 6 million tonnes, or by 2.1 thousand million EUA

Finally, this saving would mean an additional 40 000 jobs in the paper production cycle (based on a yield of 150 t/yr per person employed.

Waste paper's contribution to this saving would be about half (i.e. 3 million tonnes) raising the Community's utilisation rate to around 52 per cent, which is in fact the rate achieved by Denmark, the Netherlands and the United Kingdom. The 9 - 10 million tonnes of waste paper used at present correspond to a utilisation rate of about 43 per cent (compared with a production figure of 23 million tonnes). With a utilisation rate of 52 per cent, 12 million tonnes of waste paper would currently be used. By 1995, Community production will be roughly 30 million t/yr.

We should realise that for a whole series of technical reasons (contaminants, recycling collapse, etc.), it is unlikely that more than 60 per cent of recycled fibres will be used to manufacture paper. (This would mean 3.7 million tonnes over and above the 9 - 10 million currently recycled).

To achieve a reasonable development target of 6 million t/yr would mean maintaining the waste paper utilisation rate in conventional sorts of packaging (corrugated board, solid fibreboard etc.) while raising that in the manufacture of printing papers, including newsprint (currently 6 per cent), to somewhere between 25 and 30 per cent.

Assuming that the higher rate is possible, the quantity of waste paper used by Community industry would go up from 15 to 16 million t/yr; with an estimated consumption in 1995 of about 40 million tonnes thus giving a recovery rate of approximately 40 per cent compared with 32 per cent today.

Wrapping paper and board

The utilisation of wrapping paper and board is between 70 and 75 per cent, which is near the 75 - 80 per cent limit set by the fact that the same fibres cannot be recycled indefinitely.

In future, the quantity of waste paper used for these kinds of paper and board will depend on the quantity of finished paper and board manufactured in the Community. At the moment this is roughly 12 million t/yr and could reach 16 million t/yr within fifteen years, increasing the weight of waste paper used by 3 million t/yr (4 million tonnes x 0.75, being the optimum level).

With household and hygiene papers the waste paper utilisation rate is roughly 35 per cent. This could increase, but as the quantity produced in the Community is relatively small (1 million t/yr at present), a higher utilisation rate will have a slight effect only on the total amount of waste paper consumed.

Even if the quantity of household and hygiene papers were to double in the next fifteen years to reach 2 million t/yr , a waste paper utilisation rate of 50 per cent would result only in a quantitative increase in waste paper of about 600 000 t/yr.

Worth noting, however, is that this quantity of waste paper would replace chemical pulp, of which the Community imports 80 per cent of its needs.

It is in the printing paper sector that the increase in the waste paper utilisation rate would be really significant. We have already seen that this rate could rise to as much as 25 per cent (against 6 per cent today).

This could be achieved partly by developing Community production capacity based on a greater utilisation of waste paper and replacing certain types of paper which are currently imported in large quantities.

This applies for example to newsprint, for which community production is 1.6 million and imports are 2.5 million tonnes/year.

If the waste paper reutilisation rate were increased from today's 20 to 25 per cent to a reasonable 40 per cent, this would result in a further 700 000 t/year of waste paper being used.

The same argument can be applied to other printing papers, where one can forecast an increase in consumption and, consequently, an increase of about 2.5 million t/yr in Community production over the next fifteen years, to be achieved in part by using 700 000 tonnes of waste paper a year.

Altogether in respect of the last two sectors,
     700 000 t/yr for newsprint,
     700 000 t/yr for other printing papers, and
     600 000 t/yr for household and hygiene papers
would give us increased paper consumption of roughly 2 million t/yr. These estimates are the result of increasing both the waste paper utilisation rate in these sectors and production levels.

It must be realised, however, that generally waste paper would replace mechanical wood pulp, which is essentially wood fibres of Community origin. This aspect of the problem is obviously important.

However, wood is a raw material with many uses : building materials, sources of fibre, chemical and energy products. All these properties are insufficiently exploited in the Community today, and research really should be carried out in this field to make it possible, in particular, to improve the manufacture of chemical pulp in the long term.

Similarly, mechanically produced pulp could gradually be replaced by waste paper, and pulp wood would be better used by manufacturing pulp chemically. Dream or reality ? Only the future can say. Both the environment and the economy would benefit considerably from these changes.

Let us now consider the main ways of implementing such a policy.

Many of the measures in the field of waste paper, such as price stabili-sation, concern local authorities rather than the European Community. The Commission's role is primarily one of encouragement and coordination. However, the following priority measures are to be considered :

    a.   encouraging the public services to use recycled paper :
    b.   de-inking and decontamination treatment techniques should
          be developed to enable paper products which contain a certain
          proportion of recycled fibres to compete, on both quality
          and price, with paper manufactured from virgin pulp.
    c.   specifications for the various paper products should be
          drawn up to enable the quality of manufactured paper to be
          better suited to their use and to avoid the presence of
          contaminated substances which would preclude or impede re-
          clying operations.

In addition the public should be made aware of the advantages and the availability of products made from recycled paper.

These steps are approved by the Waste Management Committee and by the Council of Environment Ministers at its meeting on 9 April 1979. A request was made that emphasis should be given in particular to the demand for waste paper.

Taking the above guidelines and prospects for the development of waste paper utilisation as a starting point, the Commission has had two studies carried out on specifications for various paper products.

The first of these examines the use-related optimum quality standards which a paper product should meet. The strict observances of these could

enable more waste paper to be used and make it possible to reduce investment and operating costs, as certain stages in the production process
(e.g. bleaching) would become superfluous.

The second study being carried out concerns the nature and volume of
contaminants in paper, such as extenders and additives, adhesives, and
even tars. It looks into whether they are required for production,
their effect on the quality of the fibre and finished product, whether
they can be dispensed with, replaced by less harmful contaminants and
possibly prohibited, and the consequences which might result from such
measures for the production of paper and board.

Both studies will be finished by the end of this year. The Commission
will then see what conclusions are to be drawn from them.

Finally, the Commission has drawn up a recommendation to Member States
asking them to prepare and implement policies which will promote the
utilisation of recycled paper and board and, in particular :

- to encourage the use of recycled - and recyclable - paper
  and board, especially by the national government, public
  authorities and public services, all of which can set a good
  example ;
- to encourage as much as possible the utilisation of recycled
  paper and board containing a high percentage of low quality
  waste paper ;
- to reconsider, in the light of recent technological progress,
  the current requirements for paper based products which, for
  reasons other than the suitability of a product for a particular use, are prejudicial to the use of recycled paper ;
- to introduce schemes for educating consumers and manufacturers so that they will promote paper and board products
  made from recycled fibres ;
- to develop and promote the use of waste paper for purposes
  other than the manufacture of paper and board.

WASTE PAPER USE IN THE COMMUNITY

AS SEEN BY THE PAPER INDUSTRY

Dr George Holzhey

Cepac Waste Paper Committee

SYNOPSIS

Only some 75 per cent of the Community's paper consumption is covered by its own paper industry, which is heavily dependent on imported raw and semi-processed materials (roughly 1/3). Waste paper is by far and away (57 per cent) the most important "indigenous" raw material. The use of 10.1 million tonnes of waste paper in the production of 23.3 million tonnes of paper means a use rate of 43.1 per cent.

During the last 30 years the paper industry has increase its waste paper use rate from 27 per cent to 43.1 per cent. In absolute figures this means an increase of over 400 per cent.

This follows successful development of suitable machines and technologies.

Factor of international competition quality control is important in EEC countries.

Only limited quantities of good quality waste paper are available, while heterogenous and less clean waste paper is generally available in large quantities.

The price charged for waste paper is determined by the price of primary raw or semi-processed materials (wood, chemical or mechanical wood pulp), with all the costs involved in manufacturing and processing these rival fibrous materials having to be taken into account.

In view of the glaring differences in the use of waste paper in the making of various qualities of paper and board, it is easy to see that the structure of a country's paper industry is very important in making a fair assessment of that country's waste paper use rate.

RESUME

Soixante quinze pour cent suelement de la consommation de papier de la
CEE sont couverts par la production communautaire qui est, pour sa part
tributaire dans une large mesure des importations de matières brutes et
demi-produits (un tiers environ). Les vieux papiers sont de loin (57 %)
la matière première "indigène" la plus importante de notre industrie
(voir tableau 1). Avec 10,1 millions de tonnes de vieux papiers utilisés
pour la production de 23,3 million de tonnes de papier, on obtient un
taux d'utilisation de 43,1 %.

A cours des 30 dernières années notre industrie a porte le taux d'utilisation de 27 a 43,1 %; en chiffres absolus, cela représente une
augmentation de 400 %.

Le développement difficile de machines et des technologies appropriées
ont permis d'utiliser de telles quantités de vieux papiers pour des produits
finis compétitifs, qui sont fabriqués à l'étranger essentiellement à
partir de la matière première primaire abondante.

Le prix applicable pour les vieux papiers est fonction du prix des
produits de base ou demi-produits primaires (bois - pâte chimique - pâte
mécanique), etant entendu qu'il faut tenir compte de tous le coûts
supportés pour la fabrication et le traitement de ces matières fibreuses
concurrentes.

Compte tenu des différences dans l'utilisation des vieux papiers selon
les catégories de papier et de carton on comprend facilement que la
structure de l'industrie papetière d'un pays soit un facteur important
pour pouvoir porter une juste appréciation sur le taux d'utilisation des
vieux papiers dans ce pays.

Zusammenfassung

Nur rund 75 % des Papierverbrauchs der EG werden aus eigener Produktion
gedeckt, die ihrerseits in starken Masse auf importe von Rohund
Halbstoffen (ca. 1/3) angewiesen ist. Altpapier ist mit grossem Abstand
(57 %) der bedeutendste "einheimische" Rohstoff unserer Industrie. Aus
den für die Produktion von 23,3 Mio to eingesetzten 10,1 to Altpapier
ergibt sich eine Einsatzquote von 43,1 %.

In den letzten 30 Jahren die Einsatzquote von 27 % auf 43,1 % steigerte; in absoluter Menge bedeutet dies eine Anhebung um über 400 %.

Das Cerdienst der heutigen hohen Altpapier-Einsatzquote dürfen sich die Papierfabriken, die Maschinenbaufabriken und die Chemikalienhersteller gemiensam zuschreiben, denn nur die mühsame und schliesslich erfolgreiche Entwicklung geeigneter Maschinen und Technologien ermöglichte den Einsatz so hoher Altpapiermengen für konkurrenzfähige Fertigprodukte, die im Ausland vorwiegend mit problemlosen Primärrohstoffen hergestellt werden.

Gute Altpapier-Qualitaten sind nur begrenzt verfügbar. Weniger sauberes und wenniger sortenreines Altpapier (gemischtes Altpapier) hingegon ist in der Regel in grossen Mengen verfügbar.

Der für Altpapier anlegbare Preis orientiert sich am Preis der Primär-roh-oder-halbstoffe (Holz-Zellstoff-Holzstoff), wobei alle bei der Herstellung und Aufbereitung dieser dieser untereinander konkurrierenden Faserstoffe entstehenden Kosten mitberücksichtigt werden müssen.

Wegen eklatanter Unterschiede des Altpapier-Einsatzes bei verschiedenen Papier- und Pappesorten, ist es leicht erklärbar, dass die Struktur der Papierindustrie eines Landes von grosser Wichtigkeit is, um die Altpapier-Einsatzquote dieses Landes richtig beurteilen zu können.

Paper consumption in the EEC is covered only up to about 75 per cent by its own production, and production in turn is strongly dependent on imports of raw materials and semi-finished products (about 1/3). (See Table I). Among the "domestic" raw materials (produced in the EEC) waste paper has by the far the largest share with about 60 per cent.

The figures show on the one hand that the EEC paper industry is clearly in confrontation with foreign competition, and on the other hand that waste paper has a major importance as a raw material for the industry.

As regards competition from outside the EEC countries, the notable supplying countries in Northern Europe and North America draw considerable advantages from their resources of wood and energy, and they use considerably less waste paper.

From Table I we can also draw the utilisation rate of waste paper in the EEC paper industry. In a production of 24.9 million tonnes, 10.6 million tonnes of waste paper were incorporated, which gives a utilisation rate of 42.6 per cent. After Japan, this is the highest rate in the world.

Chart no. 2 shows that the paper industry can refer to respectable increases in waste paper utilisation over the past 30 years. From 27 per cent to 43.1 per cent in 1978 is a growth of 60 per cent. In absolute quantities, waste paper utilisation today is 450 per cent higher than in 1950.

In fact the industry has been using waste paper at least since 1366, the date of the oldest indication of waste paper utilisation. On the industrial level, waste paper has been systematically used since the 20s. The utilisation rate of 40 per cent set up by the EEC Commission in 1973 for 1980 has long ago been achieved and overstepped.

Commendation for the high utilisation rate is due to the paper mills, the machine manufacturers and to the chemical industry. This is because it was the difficult but finally successful development of adequate machinery and technologies that allow the use of such large quantities of waste paper in the manufacturing of competitive finished products.

In fact the EEC paper industry as a whole has achieved a leading position with its contribution to the protection of the environment.

Chart no. 3 shows that waste paper utilisation differs strongly from Member State to Member State, for instance from 18 to 59 per cent in 1976.

Why are there such big differences ? Is it right to give Belgium the
worst mark, Denmark the highest and Germany an average one ?

In order to be aware of the factors needed for a reasonable assessment of
the situation, it is necessary to first of all have a general idea of the
influences to be taken into consideration in rational utilisation of waste
paper for the manufacturing of a new grade of paper.

At the waste paper level, the following factors are important :
- quality requirements of the paper and board to be produced ;
- waste paper conditioning techniques and technology ;
- influence of international competition on finished products ;
- availability and cost of virgin fibres.

At the level of waste paper availability, the following factors are
important :
- grades and quality of waste paper ;
- availability of the required waste paper grades ;
- cost, cost fluctuation and competitiveness compared to virgin
  fibres.

These factors have a different impact depending on the country and
especially the paper grade concerned. Furthermore the structure of the
paper industry of a country has a major influence on its waste paper
utilisation rate.

The most important criterium for the waste paper utilisation potential is
the requirement of the market concerning the finished paper grades. The
industrial client does not ask for the recipe of the manufacturing of the
paper, but requires a number of specific qualities from this paper.
Whiteness, resistance, printability, smoothness, dimensional stability
and decay resistance are a few of these required properties. To a large
extent they are technically conditioned and to some extent they refer to
customer habits.

The housewife, not the papermaker, decides whether the manufacturing of
toilet paper will be done on the basis of virgin pulp fibres or recycled
fibres, because she is the one who chooses between soft white tissue paper
or grey and rough creped paper.

Resistance can only be obtained with waste paper containing a high amount
of chemical woodpulp and white printing papers can only be made from white

or easilyde-inkable waste paper. For wrapping papers for foodstuffs it is important that the waste paper does not contain toxic substances. The matter is complex because there are over 3 000 different paper and board grades, each with its particular characteristic.

In order to meet quality requirements when utilising waste paper, an adequate manufacturing technology is needed.

The growing use of upgrading processes for paper and board products, the variety of by-materials entering production and converting, as well as the constant change of their composition and characteristics on the one hand, the increased heterogeneity of waste paper grades and the continued reduction of the quality of grading, on the other hand, require that processing techniques and technologies be adjusted and renewed. An increase of the use of waste paper for graphic papers, for instance, is only made possible through a continued development of deinking and grading technologies.

Different regulatory provisions, i.e. environmental regulations, are also likely to slow down the development or the use of some processing techniques for waste paper.

Another interesting factor when assessing the possibilities to use waste paper is the extent of the international competition that a given paper or board product or a national paper and board industry has to face. As imports into the EEC cover one third of the paper and board consumption, the Community manufacturer must be very careful not to give ground, to predominantly virgin fibres based imports. For some grades, the import rate exceeds 30 per cent and in such cases (i.e. newsprint), the use of waste paper raises particularly acute problems.

The competitive behaviour of virgin fibres as compared with secondary fibres also plays a role within the EEC. The national wood supply varies very much between countries; countries with a major shortage of wood are forced to import high priced virgin fibres and, as a consequence, they are more dependent on large quantities of waste paper to be found on the domestic market. In countries where the wood supply is sufficient, the compulsion to do so is less pressing and the interest of forest owners and saw mills for "wood waste" just as strong as the interest of municipal authorities in using paper waste. It is thus incumbent on the authorities to strike a fair balance between the different interests involved.

The paper and board industry oppose fiercely the idea that some paper grades, that is some paper and board manufacturers, should be considered to be non-polluting. Paper is by all means non polluting, as it decomposes easily, it may be dissolved and reintroduced into the economic circuit. This also applies to paper grades without secondary fibres, which are precisely necessary to qualitatively rejuvenate the paper circuit with virgin fibres. Otherwise, an extensive use of waste paper is simply not possible in countries deprived of virgin fibres.

Meaningless claims about the many trees to be saved through using waste paper should be forgotten. The paper and board industry uses only forest and saw mill waste and to that extent greatly contributes to the quality of the environment and to a fair waste recycling.

The concept of "quality" for waste paper refers to the type of fibres it contains (chemical pulp, mechanical pulp, short fibres, long fibres), to the colour, to the quantity of by-materials, to the filler and the coatings, to the homogeneity of the material (purity of the qualities), to the type and the extent of the printing, to the degree of purity and to the quantity of improper materials.

As a general principle, the more homogeneous and pure the grade of waste paper is, the more diversified and effective the possibilities or re-utilisation are. The fewer processing stages there are between the manu-facturing of the paper and the final user, the less time goes by between collecting and re-utilising waste paper, the less danger there is that the waste paper will get dirty and the purer its quality will be.

The more extensive the use of waste paper is in a given country and the smaller virgin pulp based paper and board imports are in that country, the poorer the quality of waste paper will be, as it is frequently recycled, while the possibility of regenerating it with virgin fibres is restricted.

The conditions prevailing in a country also have a major influence on availability : this refers, among other things, to the density of the population, the standard of living, the customs and habits and the extent of industrialisation. Densely population countries such as the Netherlands allow for an easier recovery of waste paper than low density countries such as Scandinavian countries. In the same way, the standard of living,

which is itself correlated with a large consumption and with a sophisticated and extensive production and conversion of paper and board, is
propitious to an extensive recovery of waste paper.

These conditions also strongly influence the cost of recovering waste
paper and thus the price of waste paper. A concentrated collection and
limited transportation distances to the paper mills are major conditions
for the competitiveness of waste paper as a raw material. Only an
advantageous price will encourage the pulp manufacturer to use waste paper.

As wood supply is very limited in the EEC countries and as the production
of chemical pulp does not meet the specific needs, waste paper is very
important as a substitute for pulp. The collection and the use of the
qualities of waste paper with a high content of chemical pulp are
intensive; the demand for these qualities is strong and their prices
follow those of chemical pulps.

In other words well graded waste paper containing chemical pulp is always
in demand provided that it is priced under virgin pulp. However, medium,
lower and brown qualities of waste paper, which also contain chemical
pulp fibres and which may be used for the manufacturing of wrapping paper,
cannot align their prices with that of bleached Kraft pulp, but rather
they must follow the import prices of Kraft papers, which are called
Kraftliner. This hierarchy explains the wide price and usage fluctuations
of waste paper.

Kraftliner production in North America is both extensive (about 18 million
t/year), and cost competitive. An economic slow down in the U.S. reduces
home consumption and forces part of the production on the export market
at discounted prices. The European competitors attempt to maintain their
market share in a downward pressure on the prices of waste paper. The
process is inverted with an economic upturn.

As wrapping papers and corrugated paper based papers use up about 85 per
cent of all the waste paper available, these developments also influence
the other waste paper qualities.

These fluctuations of waste paper prices limit its use in the manufacturing
of paper qualities for which waste paper competes with mechanical wood
pulp and for which the use of waste paper requires major investments.
The de-inking technique has developed substantially, but such facilities
require a big capital outlay.

Finally, the structure of a national paper and board industry is important for the utilisation rate of waste paper in that country (Table 4). The larger use is for wrapping papers (approximately 75 per cent). For corrugating papers, it even reaches 84 per cent. As a whole, 8.2 mio t are used for wrapping papers, which represents 87 per cent of the waste paper consumption in the EEC. For printing/writing papers, that is the second larger group, with 43 per cent of production, only 550 000 t of waste paper are used, that is 5.8 per cent of the waste paper consumption in the EEC. The toilet and special papers group is not very important, in spite of relatively high rates of utilisation, as it only represents only 7.5 per cent of the EEC paper and board production.

Because of these enormous differences in the use of waste paper for the different grades of paper and board, it is easy to explain that the over-all rate of utilisation of waste paper varies according to the structure of the national paper and board industries.

A country as Belgium, for instance, with 65 per cent of its overall production consisting of printing/writing papers and 6 per cent for food-stuffs wrapping papers, for which the use of waste paper is governed by stringent regulations, should normally have a rate of utilisation well under the Community average.

On the contrary, Denmark, thanks to a small quantity of printing/writing papers, can easily reach a rate of utilisation of waste paper of 50 to 60 per cent. In Great Britain and in the Netherlands, where the share of printing/writing papers is under the EEC average (35 - 40 per cent), the rate of utilisation of waste paper is above the EEC average.

The industry is willing to discuss reasonable actions aimed at increasing waste paper recycling. It is crucially interested in this because of the limited wood resources of the EEC means that capacity increase is only possible if the use of waste paper increases.

However, authorities must be warned away from measures tending to influence the freedom of choice of raw materials. The Federal Republic of Germany has taken account of these basic principles with good results. It could be followed by the EEC or by other countries.

High usage of waste paper requires a diversified and highly productive national paper and board industry ... it should be internationally competitive ... it should be able to adjust swiftly to the changes in market requirements and manufacturing techniques.

The industry supports assistance to research and development programmes, as well as to efforts to improve selective waste collections from households. Fair measures to encourage the provisional storage of the excess waste paper quantities during a slowdown of the economy are certainly helpful for the industry.

In answer to a question on the affect of the value of the dollar on the competition between virgin and secondary fibres, if the dollar depreciates further, the impact on the use of waste paper would without any doubt be negative. This is because virgin fibres could be imported at low costs in the form of pulp and because the finished products of the dollar area would become more competitive.

STRUCTURE OF THE WASTE PAPER PROCESSING INDUSTRY
IN WESTERN EUROPE

Ben van der Weerden

S.Levison B.V.

The Netherlands

SYNOPSIS

INTRODUCTION

The speaker dismisses a few existing myths about waste paper and goes on to make several possible subdivisions of waste paper.

1.   Collection of Waste Paper and Its Costs

a.   Collection of waste paper
Industrially produced waste paper has great influence on the market.

b.   Vulnerability of the collection system
Per capita consumption and the control of the waste paper market in situations of excess demand and supply.

c.   Costs of collection, sorting and bailing
High costs illustrated by Dutch and German studies, cost per tonne being £26.43 and £30.63 respectively.

2.   The Structure of the Waste Paper Processing Industry and the Integration with Paper and Board Mills

a.   The structure of the waste paper processing industry
This has the form of a pyramid divided into three tiers :
1.   Hawkers, volunteers, local authorities and small companies.
2.   Locally orientated companies with baling facilities.
3.   Mill suppliers with wholesale functions.

b.   Integration
The speaker discusses the reasons and the pros and cons of integration from the point of view of the paper and board mills and the waste paper companies respectively. Furthermore, the three main working forms of integration which occur in Europe are dealt with and finally it is put that the consequences of integration have led to a

more stable and transparent market.

3. *The Role of Legislation*

According to Mr Van der Weerden the use and the quantative collection of waste paper is solely governed by demand and not by legislation from the "authorities". Also the actual role of local authorities in the collection of waste paper endangers the unique collection-system of waste paper by the waste paper dealers, which system has enabled the paper- and board industry to always receive sufficient raw material against reasonable prices.

If legislation has to be carried out it has to be done in cooperation with the waste paper branch, otherwise it could have disasterous effects.

Nevertheless, if governments keep on interferring in the market they should accept the consequences, being a risky and very expensive waste paper fund.

RESUME

Introduction

L'orateur detruit certains préjugés en ce qui concerne les vieux papiers et distingue plusieurs catégories.

1. *La Collecte des Vieux Papiers et son Coût*

a) *La collecte des vieux papiers*
   Les vieux papiers d'origine industrielle ont une forte incidence sur le marché.

b) *Vulnérabilité du système de collecte*
   La consommation par tête et le contrôle du marché des vieux papiers en cas d'excès de l'offre et de la demande.

c) *Coûts de la collecte, du tri et de la mise en balles*
   Coûts élevés d'après les études néerlandaises et allemandes, le coût par tonne étant respectivement de $26,43 et de 30,63t.

2. *La stucture de l'industrie du recyclage des vieux papiers et son intégration avec les usines de papier et de carton*

a) *La structure de l'industrie du recyclage des vieux papiers*
   Elle se présente sous la forme d'une pyramide à trois étages :

1. récupérateurs, volontaires, autorités locales, petites sociétés;

2.   entreprises axées sur le marché local disposant d'installations de mise en balles;

3.   fournisseurs des usines de papeterie faisant offre de grossistes.

b)   Intégration

L'orateur examine les raisons et les avantages et inconvénients de l'intégration tant du point de vue des fabricants de carton et de papier que de celui des enterprises de récupération et de recyclage des vieux papiers. It traite en outre des trois principales formes d'intégration que l'on trouve en Europe et en tire finalement la conclusion que l'intégration s'est traduite par une plus grande stabilité en transparence du marché.

3.   Le rôle de la Législation

Selon M. van der Weerden, l'utilisation et le ramassage des vieux papiers ne sont fonction que de la demande et ne dépendent pas de la législation arrêtée par les "autorités". En outre, le rôle joué actuellement par les autorités locales dans le ramassage des vieux papiers compromet le système unique de ramassage des vieux papiers mis en place par le négoce, système qui a toujours permis à l'industrie du carton et du papier de se procurer des matières premières en quantités suffisantes et à des prix raisonnables.

Si une législation doit être mise en place, elle devra être élaborée en coopération avec le secteur des vieux papiers, faute de quoi elle ne pourrait avoir que des effets désastreux. Cependant, si les gouvernements continuent à intervenir sur le marché, ils devront en accepter les conséquences et créer un fonds, ce qui serait une opération aléatoire et très coûteuse.

ZUSAMMENFASSUNG

Einleitung

Der Vortragende räumt mit einigen Mythen über Altpapier auf und nimmt mehrere mögliche Unterteilungen von Altpapier vor.

1.   Einsammeln von Altpapier und Seine Kosten

a)   Einsammeln von Altpapier.
Industriell erzeugtes Altpapier beeinflusst den Markt in starkem Masse.

b)   Schwachstellen des Sammelsystems

Pro-Kopf-Verbrauch und Kontrolle des Altpapiermarktes in Situationen übermässiger Nachfrage und Belieferung.

c)   Kosten für das Einsammeln, Sortieren und Pressen
Durch niederländische und deutsche Studien nachgewiesene hohe
Kosten die sich auf £26.43 beziehungsweise £30.63 pro Tonne
belaufen.

2.   *Die Struktur der Altpapierverarbeitungsindustrie und die
Integration mit Papier-und Pappefabriken*

a)   Struktur der Altpapierverarbeitungsindustrie
Sid weist die Form einer in drei Abschnitte unterteilten Pyramide
auf.

1.   Strassenhändler, Freiwillige, Ortsbehörden und Klein unternehmen.

2.   Örtlich ausgerichtete Unternehmen mit Pressanlagen.

3.   Papierfabrikl eferanten mit Grosshandelsfunktionen.

b)   Integration
Der Redner erörtert die Gründe und das Für und Wider der Integration
vom Standpunkt der Papier- und Pappefabriken beziehungsweise der
Altpapierunternehmen. Ferner werden die drei wichtigsten praktischen
Formen der Integration, die wir in Europa finden, behandelt;
schliesslich wird ausgeführt, dass die Integration zu einem
stabileren und transparenteren Markt beigetragen hat.

3.   *Rolle der Gesetzgebung*
Nach den Ausführungen von Herrn van der Weerden ist für die
Verwendung und das quantitative Einsammeln von Altpapier ausschliess-
lich die Nachfrage ausschlaggebend und nicht die Gesetzgebung der
"Behörden". Die eigentliche Rolle der Ortsbehörden beim Einsammeln
von Altpapier gefährdet im übrigen das einzigartige Sammelsystem
von Altpapier durch die Altpapierhändler; aufgrund dieses Systems
konnte die Papier- und Pappeindustrie jederzeit ausreichend
Rohmaterial zu vertretbaren Preisen beschaffen.

Die Durchführung eines Gesetzes muss in Zusammenarbeit mit der
Altpapierbranche erfolgen, da sich anderenfalls katastrophale
Auswirkungen er geben könnten. Falls jedoch die Regierung
weiterhin in den Markt eingreift, muss sie auch die Folgen tragen
nämlich Errichtung eines risikobehafteten und sehr kostspieligen
Altpapierfonds.

One of the biggest handicaps faced by the waste paper trade in policy
making is that too many people consider themselves experts in the field.
A few myths[*] to be done away with, once and for all, are :

1. Waste paper is free. It is not as soon as it starts being
    accumulated in large quantities it picks up cost.
2. Waste paper is newspaper alone. Newspaper accounts for less
    than a fifth of the total.
3. Recycling can be increased by collecting more waste paper.
    This is not true. It is not the insistence of society that
    determines the demand for waste paper, but the economics of
    papermaking.
4. The waste paper all around us is immediately available for
    use. It is not. Shutting the flow off, or starting it, will
    cause problems, for it cannot be manufactured. A wastepaper
    merchant is constantly in search of new sources and outlets.
5. The more waste paper is collected, the more will be used; or,
    the cheaper it is, the more will be collected. These statements
    are also not true. The use of waste paper is governed only by
    the number of mills that can use 100 per cent secondary fibre.

Having put the above myths to bed, precision is also necessary in cate-
gorising different kinds of waste paper. It can be subdivided in several
ways, including by :

lower grades and better grades ;
industrial waste and waste from voluntary collections ; and
de-inking and non-de-inking grades

Throughout the western world, the waste paper trade plays a dominant role
in the acquisition of waste paper. In accordance with the fifth myth, it
is the demand which always determines the amount collected.

Industrial paper waste is always collected, no matter if the waste-
producer receives money for it or has to pay to get rid of his waste paper.
Thus industrial waste paper makes up a basic quantity; it is the difference between this quantity and demand that determines waste paper
collection activity by volunteers from households.

---

[*]From : "The six myths of waste paper" by Peter Block, Communications
Manager, Consolidated Fibres Inc., San Francisco, California.

With the quantity of industrial waste paper being fixed, the quantity of waste paper from volunteers suffers in an unproportional way from market ups and downs.

The constant generation of industrial waste paper also results in wide price fluctuations. Too often prices have dropped too low for both the waste paper companies, and the voluntary groups. Decisions to stop collection have had to happen far too often, generating a feeling of untrustworthiness among the volunteers.

The importance in different countries of the quantity collected by volunteers depends mainly on the waste paper consumption per capita. A low consumption per capita generally means that the demand for waste paper can almost only be covered by industrial waste; a high waste paper consumption means that the demand for waste paper, even in times of low capacity-utilization, has to be covered to an important extent by the voluntary collections, as in the Netherlands.

A second aspect is the "fly-wheel effect" which occurs when stimulating or "curbing" voluntary collections. In situations of excess supply, the collection does not stop immediately but only decreases slowly. In situations of "excess demand" it takes a long time to convince volunteers to restart activities.

Governmental interference in the system is only useful and welcome if it takes the form of conditional financial aid during bad market conditions to help small companies survive. These companies are important, but their overheads have to be met even when waste paper from the voluntary is not in demand.

Cost of collection

Since the time of the waste paper dealer on a carrier-tricycle, the business of waste paper processing has changed enormously. Large high capacity continuously operating presses have been introduced. Transport now involves expensive container systems. Studies have been carried out in Germany ('76) and the Netherlands ('79) to reveal costs involved in collecting, sorting and bailing waste paper.

COSTING CALCULATION WASTE PAPER COMPANY NETHERLANDS

Based on 10 000 tonnes per year.

INVESTMENTS
.   Building + Land   )
.   2 bailing presses   )
.   Weighbridge   )
.   3 containers vehicles   )   Hfl. 2 130 000   £ 489 655
.   40 containers   )
.   Truck   )
.   Sundries   )

YEARLY COSTS
.   10 men   )
.   Transport costs   )
.   Factory costs   )
.   General costs   )   Hfl. 1 150 000   £ 264 368
.   Interest   )
.   Depreciation   )
.   Etc.   )

Cost per tonne $\frac{1\ 150\ 000}{10\ 000}$ = Hfl. 115.-- *   £ 26.43

*   These costs thus include : . collection   (£1 = Hfl. 4.35)
                                . sorting
                                . bailing

    But exclude   : . buying price
                               transport to mill
                               profit

COSTING CALCULATION WASTE PAPER COMPANY WEST GERMANY

Based on 9 600 tonnes per year.

VARIABLE COSTS

.	Energy	)		
.	Oil	)	DM 96 000	£ 24 303

FIXED COSTS

.	Personnel	)		
.	Factory costs	)		
.	General costs	)	DM 1 065 600	£ 269 772
.	Interest	)		
.	Depreciation	)		
.	Etc.	)		

Cost per tonne $\dfrac{1\ 161\ 600}{9\ 600}$ = DM 121.-- *   £ 30.63

*   These costs thus include : . collection
                                 . sorting
                                 . bailing

     But exclude   : . buying price
                                 . transport to mill
                                 . profit

     (£1 = DM 3.95)

Structure of the industry

Although the structure of our branch of industry is not exactly the same in all countries of the European Community and although there are some extreme deviations, a same basic pattern exists in all the member countries.

The structure of the waste paper processing industry, which varies from country to country, usually takes the form of a pyramid. At the base are hawkers, volunteer groups and in some instances local authorities.

Hawkers are small one-man waste paper dealers who only buy and pick up the waste paper from various sources.

At the same level in the hierarchy are groups of volunteers, who collect waste paper in order to raise money for their club, school or institution; and the local authorities that collect waste paper from door to door, a system which only results in high losses when one looks at it economically.

The second tier of the pyramid are the many locally orientated waste paper companies spread all over the country. These pick up waste paper regularly from industrial sources, department stores, supermarkets and from voluntary groups such as schools, sportclubs, churches etc.

It is on this level that the most important work has to be done and the highest costs incurred. They are often small companies, but they have to supply enormous services and facilities in order to transport the waste paper in a loose form, sort, grade and bail it.

The third and highest tier in the supply pyramid consists of the mill suppliers. These are usually bigger companies, in the wholesale trade, and they usually have one or more waste paper processing plants of their own. In some cases they have large specialized sorting plants. Their activities include : distribution, imports and exports to help solve excess demand or excess supply, financing dealers in the second tier of the pyramid, and being continuous buyers from dealers of the second tier, thus smoothing market fluctuations.

Integration into the waste paper market has progressed with the increase of consumption during the last 15 years. This includes vertical integration, when paper or board mills acquire financial interest in waste paper companies. In addition there is horizontal integration, when mills and waste paper companies form groups for supplying and buying waste paper.

Paper or board mills adopt integration in order to improve their supply of waste paper and to be able to influence the market. Vertical integration as a means of improving profitability of a waste paper company is rare.

For the waste paper company integration helps to maintain continuity of a family business sold on the retirement of the owner, aids modernisation and mechanisation, and results in a regular sales outlet.

Legislation

In the opinion of the trade, the influence of the various authorities on waste paper recovery is too orientated in favour of the ethical motive to collect as much waste paper as possible. This can be highly un-ethical, as it can destroy a unique collection system set up by many groups of volunteers. The collection system is vulnerable : any control from authority must follow collaboration with the waste paper trade.

The main goal of any legislation should be to increase consumption of those waste paper grades for which the availability can be increased by more intensive collection.

If national and supra-national governments keep on interfering with the waste paper market, they should take the responsibility of being involved, and should set up a series of large sheds all over the EEC in order to stock excess supply and to maintain a minimum price.

As no government is likely to risk such a project, the waste paper industry should be allowed to do what it has done as long as it has existed, that is to maintain a sufficient supply of waste paper.

OTHER WASTES AND ASPECTS OF WASTE-DISPOSAL POLICY

IN THE EUROPEAN ECONOMIC COMMUNITY

Dr Benno W.K.Risch

Environment and Consumer Protection Service

Commission of the European Communities

SYNOPSIS

The waste management activities of the Community were essentially concentrated until now on the general organization and the harmonization of legal instruments of waste disposal as well as on wastes representing particular hazards for the environment.

The future waste management policy of the Community will however lay more emphasis on the recovery and the reuse of wastes. Waste management shall become a fundamental instrument of a European resources policy. The policy will be extended to new areas, comprising : demolition waste ; old tyres ; sludges ; textile wastes ; ferrous scrap ; non-ferrous metal wastes ; mining wastes and cinders from power stations.

Furthermore, Community actions will concentrate on some additional priorities in order to create a certain number of technical, legal and economic framework conditions such as waste exchanges, waste data banks, R & D programmes, consumer information, promotion of the use of recycled products by the public sector, voluntary arguments, the promotion of large scale centres for municipal waste recycling.

The generation of demolition waste and in particular of concrete waste will increase in the future. The Commission is to therefore elaborate special guidelines to deal with the problem.

Old tyres : the Commission is preparing proposals in order to promote the collection and the recovery of this valuable material.

Waste oils : the Commission is preparing a directive concerning the prevention of waste oil burning without pre-treatment and without any energy recovery.

Need to promote recovery possibilities of sludge outside agriculture are
also needed for R & D actions to reduce the deterioration of ferrous scrap
reclaimed from municipal work by non-ferrous metal impurities.

Because of the high raw material and energy value of non-ferrous metal
wastes, the Commission intends to take new initiatives in this sector.
As a first step, the Commission is preparing a special R & D programme.

Increased use of coal for energy production will lead to the increase of
wastes from coal mining and of power station ashes. The Commission intends
to undertake actions in order to promote the recovery of textile wastes
with the Community. The activities of waste exchanges should be promoted
as well as their cooperation within the Community. Urgent need for waste
data banks : the Commission has therefore proposed the setting up of a
Community wide waste management data bank system.

RESUME

Autres déchets et aspects de la gestion des déchets dans la Communauté
Européenne

Les activités de le Communauté en matière de gestion des déchets étaient
concentrées jusqu'ici à l'organisation générale et l'harmonisation des
dispositions règlementaires de l'élimination des déchets ainsi qu'aux
catégories de déchets représentant des risques particuliers pour
l'environnement.

La future politique communautaires en matière de gestion des déchets
mettra l'accent plutôt sur la récupération et le réutilisation des
déchets. La gestion des déchets deviendra un élément fondamental d'une
politique communautaire des ressources.

Les travaux de la Communauté dans la domaine traités jusqu'ici comme
les déchets toxiques et dangereux, les emballages de boisson, les vieux
papiers, les huiles usagées et les déchets agricoles seront poursuivis
mais concentrés plutôt sur la promotion de la récupération de ces déchets.

Les activités seront cependant étendues aussi à d'autres domaines, comme

- les déchets de démolition
- les déchets pneumatiques
- les boues

- les déchets de textiles
- les déchets ferreux
- les déchets de métaux non ferreux
- les déchets miniers et les cendres de combustibles.

En plus, les activités futures de la Communauté seron.t concentrées à un certain nombre de domaines prioritaires en vue de créer certaines conditions fondamentales d'ordre technique, juridique et économique : bourses des déchets, banques de données, programmes de recherches et de développement, l'information du public, la promotion de l'utilisation des produits recyclés par le secteur public, les accords de branches, la promotion de plate-formes centralisées pour la récupération des déchets urbain.

ZUSAMMENFASSUNG

Giftige und gefährliche Abfälle, die insbesindere in der Industrie anfallen, stellen eines der wichtigsten Umweltprobleme dar, und ihre Verwertung und Beseitigung ist das vordringlichste qualitative Problem der Abfallwirtschaft.

Die Europäische Gemeinschaft produziert jährlich 20 Millionen Tonnen solcher Abfälle, was 15 bis 20 % des gesamten Industrieabfalls ausmacht. Nur mittel- und langfristig ist mit einer allmählichen Verringerung dieser Menge zu rechnen, wenn sich die Industrie den Rechts- und Verwaltungsvorschriften sowie den technischen Regeln anpasst und die umfangreichen Forschungs- und Entwicklungsarbeiten Ergebnisse zeitigen.

Eine besondere Richtlinie über giftige und gefährliche Abfälle wurde verabschiedet, die von den einzelnen Mitgliedstaaten seit dem 20. März 1980 angewendet werden muss.

Aus den vorliegenden Auskünften geht hervor, dass die wichtigsten Bestimmungen der Richtlinie zumindest in sechs Mitliedstaaten bereits in Kraft sind. Da es sich hierbei jedoch lediglich um eine Rahmenricht-linie handelt, muss diese Richtlinie noch ergänzt und erweitert werden.

Für die zweite Phase der Gemeinschaftsarbeiten im Bereich giftiger und gefährlicher Abfälle hat die Kommission ein mittelfristiges Arbeitsprogramm

erstellt, das aus zwei Teilen besteht: "Vorrangige Aufgaben und Massnahmen"
und "sonstige Probleme".

Nach Auffassung der Kommission sollte die Rictlinie durch Gemein-
schaftsregeln und -bestimmungen in folgenden Bereichen ergänzt werden :

- die Auserbeitung einer Rahmenrichtlinie über die Lagerung und Ent-

  lagerung giftiger und gafährlicher Abfälle, einschliesslich einheitlicher Kriterien und Regeln hinsichtlich der Auswahl, Kontrolle
  und Pflege der Deponien für derartige Abfälle;

- die Auserbeitung von Regeln über die einheitliche Bezeichnung von
  giftigen und gefährlichen Abfällen und ihrer Behälter sowie Einführung einheitlicher Gefahrensymbole;

- die Ausarbeitung einer Durchfürungsrichtlinie über den Urkunden-
  nachweis im Zusammenhang mit giftigen und gefährlichen Abfällen, in
  der insbesondere das Verzeichnis, die Indentifizierung und die
  Begleitdokumente für den Transport solcher Abfälle und vor allem
  die grenzüberschreitende Beförderung geregelt werden;

- besondere Regeln für Sicherheitsvorkehrungen bei Unfallgefahr und
  dergleichen;

- die Errichtung einer Datenbank für giftige und gefährliche Abfälle.

General : basic concepts

In the next few years, waste disposal will have to be developed into an integral part of a European resources policy. The emphasis must be on reducing and using waste while ensuring the non-polluting disposal of unavoidable, useless waste.

Most of the 2 000 million tonnes of waste produced annually in the EEC contains potentially valuable substances. The percentages and the individual quantities are considerable and of great importance both for raw materials and energy. The Community imports 80-90 per cent of its iron, tin and zinc and almost 60 per cent of its paper and cellulose.

It is assumed nowadays that between 70 and 90 per cent of all waste and residues can be re-used in one form or another. Hitherto, however, 80-90 per cent of waste has been destroyed or simply dumped.

The Community must therefore enact economic and legal outline regulations and set clear objectives for waste-disposal policy. Attention should be concentrated on the following:

promotion of co-operation between the public sector and industry at various levels;

development of information and publicity work. Compilation of uniform waste statistics and setting-up a Community waste-disposal policy data bank are important instruments in this connection;

promotion of research and development;

increased education of consumers;

reduction of excessive quality requirements for products;

promotion of use of secondary raw materials;

promotion of voluntary agreements between waste-producers (communities, industry), scrap dealers and the processing industry.

large-scale industrial use of domestic refuse. A region with about one million inhabitants is regarded as a commercial collecting district. Plans could also be made for a European demonstration and model installation for mechanical sorting, and one or two initial model large-scale projects for a waste-disposal and raw-material recovery centre. A centre could be set up at the meeting-point between the three Länder of Maastricht, Lüttich and Aachen.

Community measures on the following kinds of waste:
demolition waste, used tyres, sludge, textile waste, waste from the
food industry and agriculture, ferrous and non-ferrous metals,
chemical waste, and waste energy.

In addition to the first-priority sectors, the Waste-Disposal Policy
Committee of the Commission has chosen as second-priority sectors:
demolition waste, used tyres, sludge and textile waste.

Demolition waste

Every year, about 160 million tonnes of demolition waste and rubble
accumulate in the EEC Member States. In coming years there will be a
disproportionate increase, partly because of building renovation and
partly because of the disintegration of concrete structures. About
10 per cent, ie about 16 million tonnes, can be used as a covering
material for dumps and other excavation work, but the remaining amount has
to be disposed of. This requires enormous dumping areas. Concrete waste,
which makes up between 40 and 50 per cent of the total, causes special
problems owing to its weight and to the fact that it comes in large lumps.
There are also a number of environmental problems (noise, pollution and
safety aspects).

More particularly, the amount of waste concrete will increase very greatly
in the next few years. It will make up more than 50 per cent of the amount
of demolition waste and rubble. Prestressed concrete causes special
problems owing to the reinforcements. The re-use of demolition waste is
important from the stand-point of ecology, the supply of additives for
concrete, and scrap and energy policy. In view of the increasing demand
for scrap it is unacceptable that, within the Community alone, about 2 or
3 million tonnes of scrap iron are lost every year in concrete, which is
taken direct from the demolition sites to the dumps and tipped.

Scrap iron is also stored energy, since it has already been heated. It is
generally assumed that about DM 150 per tonne is saved in energy if scrap
iron is used instead of iron ore.

The EEC, following Japan and the USA, is also addressing itself to the
problem of demolition waste and rubble. A group of experts from Member

States are now developing a project for an EEC research and development programme aimed at reducing and re-using concrete demolition waste. It is expected to propose a research programme for coordination and indirect action on concrete waste for submission not later than next year to the Council of Ministers as part of EEC research on raw materials. Within this programme it is also proposed to set up an EEC pilot project for the non-polluting treatment of concrete waste.

In addition to the actual research and development work, the Commission is also thinking of setting up a special working party on demolition waste and rubble to deal with the special ecological, economic and legal problems in this sector. The group will discuss the following problems, inter alia:

> guidelines for demolition work;
>
> guidelines for concrete work;
>
> guidelines for re-use of demolition waste in the building industry; and
>
> guidelines for demountable buildings, which will be planned so that they can be demolished in such a manner as to cause less pollution.

Used tyres

Only a few years ago, used tyres were regarded as a burden, difficult to dispose of, refused by many dumps, and often illegally dumped, thus disfiguring the landscape. They are now finally accepted as a valuable commodity which can be nearly 100 per cent re-used. Because of the increased price of oil and raw materials, a hunt has begun for used tyres. The relevant sectors of the used-tyre market are already competitive with regard to the best economic and technical methods of re-use of tyres.

Used tyres have a high energy value, greater than that of coal, ie 8 200 Kcal/kg compared with 7 000 Kcal/kg for coal and 9 800 Kcal/kg for heavy fuel oil. Remoulding of tyres is an intermediate stage in the re-use of tyres. Well-preserved tyres, instead of being immediately burnt, pyrolyzed or regenerated, can first be remoulded.

The two most varied methods of tyre re-use are undoubtedly the regeneration of rubber and the recovery of raw materials by pyrolysis. A turning-point is being reached here also, because of the increasing price of oil and raw materials.

At present about 1.8 million tonnes of used tyres accumulate per year in the EEC Member States. Only about 800 000 tonnes are being re-used at present. About 510 000 tonnes are remoulded, and 30-50 per cent of remoulded tyres are used for lorries.

The fact that almost 60 per cent of used tyres are still being dumped, destroyed or otherwise lost is an intolerable waste of this valuable material. The Council of Ministers for the Environment, at a meeting on 9 April 1979, called on the Commission to develop a used-tyre policy.

The Commission considers that the re-use of tyres is a very important part of waste-disposal policy. Accordingly, used-tyre policy must consist of graded measures.

Firstly, it is necessary to prevent wastage of this valuable material, so destruction of used tyres is no longer permissible. Used tyres must be systematically collected and re-used.

To counteract waste and make efficient use of tyres, it is first necessary to increase the service life of tyres. It is quite possible industrially to produce a tyre which can run for up to 160 000 km.

Tyres should of course be remoulded if suitable for this purpose. The process will be simplified if new tyres are more standardized.

The Commission considers that remoulding can be greatly increased - by a further 30 to 40 per cent. It is important, however, to take greater account of safety aspects, particularly if remoulded tyres are used more extensively for passenger vehicles. Quality control and standards must therefore be developed.

Since pyrolysis and regeneration still pose a number of technical and ecological problems, research work on used tyres should be intensified. For example the Commission, in their first research programme on the promotion of recycling of domestic and industrial waste, devoted a special chapter to research and development on the re-use of tyres. With regard to the organization required for promoting re-use of tyres, it is proposed to have a "tyre delivery point" in order to counteract wastage of used tyres, increase the service life of tyres and intensify the use of remoulded tyres.

Waste oil

Waste oil, like used tyres, is not rubbish but a valuable commodity which can be re-used almost 100 per cent. Waste oil has a high energy value and, as a secondary refined material, is a valuable raw material practically equal in quality to the primary refined material.

Various research has shown that the burning of waste oil in small furnaces results in considerable nuisance from certain solid and gaseous substances. The Commission is therefore preparing a new guideline to forbid the burning of untreated waste oil.

The Commission is also considering the extent to which the use of secondary refined substances, particularly in the public sector, can be further promoted. The growing interest in this commodity is shown by the fact that, between 30 September and 2 October, the Second European Congress on Waste Oil is very interested in waste oil, it has refrained from discussing the various aspects of waste oil disposal at this conference here in London.

Sludge

The Commission considers that sludge will in future be the number one quantitative programme on waste disposal. In view of the measures which have been taken on clean water and which will become comprehensive in the next few years, there is likely to be a considerable increase in the amount of sludge from the municipalities and from industry. These cannot all be used in agriculture, so considerable efforts will be needed to reduce the amounts of sludge. In addition, applications must be found outside agriculture.

In the next few years, the Commission must also take the initiative on ferrous and non-ferrous metal waste. As a first step, the Commission will propose a special research programme in this sector.

Textile waste

Textile waste has hitherto been largely neglected. It is assumed that 2 to 3 million tonnes of textile waste accumulate annually in the EEC, of which about 70 per cent is suitable for recycling. This corresponds to an economic value of about DM 300 to 500 million. The burning of non-reusable textiles alone would have an energy value of almost 1.5 to 2 million barrels of oil.

In order to throw some light on this sector, the Commission will undertake an initial study. It will consider which Community measures can be taken to promote the re-use of textiles and will also discuss the question of increased use of textile waste with the relevant branches of industry in the EEC.

Waste exchanges

The Commission considers waste exchanges to be excellent instruments in the service of environmental protection and the re-use of production residues. From the beginning, therefore, it has had close contact with the waste exchanges operating in EEC Member States and has improved the cooperation between these exchanges. In cooperation with the Standing Conference of EEC Chambers of Industry and Trade, it has issued a manual of practical proposals for waste exchanges, which will guide all those who intend to operate in this area. The Commission also promotes cooperation between waste exchanges across the boundaries of Member States.

Information exchange on wastes

The Commission considers that the existence of a comprehensive, central data and information system on waste disposal is an urgent, necessary instrument of an efficient waste-disposal policy, particularly in view of the complex nature of waste. The Commission therefore intends to set up a European Community Information Exchange on wastes.

It considers that such a Community information system, in the case of waste and recycling, will be of great economic advantage both the the Community and to the individual countries. With the exception of the waste-disposal data bank at the Federal Office of the Environment, Berlin, the other Member States have nothing comparable. In the view of the Commission, the lack of adequate information on waste and recycling is a bottleneck resulting in losses, wastage and excessive costs in the national economies. This year the Commission hopes to begin a five-year operational pilot phase of an Information Exchange on Wastes.

Conclusion

The Commission considers that waste-disposal policy will become increasingly important; a future-oriented waste-disposal policy will preserve

the environment and save the use of raw materials and energy. There must therefore be close cooperation between the public sector, industry and consumers.

RE-USE OF DEMOLITION RUBBLE

Carlo de Pauw

Building Industry Research Centre

Belgium

SYNOPSIS

Re-use of demolition rubble

*Large quantities of building and demolition rubble will be produced
during the coming decade. Demolition of concrete constructions gives
concrete mixed with brick/masonary rubble. Following start of concrete
constructions after 2nd World War, we now find ourselves at the foot of
a demolition waste mountain. Problem of the presence of the reinforcement. But with supply difficulties of aggregates for concrete, re-use
of demolition rubble will become a necessity. Tipping spots are also
increasingly difficult to find. Recycling offers solutions to both
problems.*

*Cooperative research venture between the Netherlands, Belgium and
Germany launched in 1977.*

*When considering re-use of demolition rubble as aggregate for recycled
concrete, then the rubble from the demolition site must first be broken
down into small pieces. The presence of reinforcement, prestressed or
not, and of impurities such as gypsum, wood plastics, etc... in the
concrete or brick/ masonry rubble makes separation techniques necessary.
Just which techniques are appropriate depends on the field of application concerned. But the production from demolition rubble of concrete
for certain uses is already technically feasible.*

*Need in the future for research to consider the existing state of
concrete technology and the regulations relating to concrete in line
with the possible substitution of traditional constituents by rubble
or other waste materials.*

RESUME

Enfoncé du problème

Il faut bien admettre que dans les décennies à venir, la rénovation,
d'une part, et la démolition de batîments, d'autre part, vont produire de
grandes quantités de déchets de démolition. Alors que la première fera
apparaître surtout des débris de natures diverses, la seconde conduira à
des déchets constitués principalement de béton mélangé plus ou moins à
autre chose, mais surtout à des débris de maçonnerie. Une chose est sûre
tôt ou tard, tout objet arrive à l'état de déchet, et il en est de même
pour une construction.

Etant donné la durée de vie moyenne des constructions en béton et le fait
que l'emploi du béton s'est développé surtout après la deuxième guerre
mondiale, on se trouve maintenant au pied d'une montagne de débris de
démolition. Parmi eux, c'est le béton armé qui pose le problème de
traitement le plus difficile, précisément à cause des armatures. Le
remploi des déchets de béton s'impose non seulement, au point de vue
écologique, mais aussi parce qu'il peut fournir des granulats pour
béton. En effet il est de plus en plus difficile de trouver des aires
de décharge et, quant aux sources de granulats naturels, elles sont soit
épuisées, soit plus difficile à exploiter à cause de certains mouvements
d'opposition. Le recyclage répond en même temps à ces deux problèmes.

Il n'y a donc rien d'étonnant si, après les Etats-Unis et le Japon,
quelqu s pays de la Communauté, parmi ceux où les espaces bâtis sont les
plus denses, ont vu s'imposer la nécessité de recycler les déchets de
construction et les débris de démolition. C'est en effet à l'initiative
de l'organisation néerlandaise de recherche CUR que, en 1977, s'est

établie une collaboration entre les Pays-Bas, la Belgique et l'Allemagne,
qui a abouti momentanément aux Pays-Bas et en Belgique à une recherche sur
le thème: demolir en respectant l'environnement, démonter et recycler le
béton, etc... et aussi remployer les débris de maçonnerie.

Ces differentes parties sont étroitement liées. En effet, le fait de
trouver des domaines d'application pour le béton recyclé peut rendre
nécessaire d'utiliser de nouvelles techniques de démolition; par exemple,
on pourrait démolir de façon sélective pour éviter les impuretés dans le

béton recyclé. D'autre part, construire "démontable" est véritablement
une nouvelle stratégie qui, dès le projet de construction, introduit une
technique de démolition qui respecte l'environnement.

Traitement des déchets de démolition

Les déchets de démolition qu'on veut remployer comme granulats dans ce
qu'on appelle du nouveau béton recyclé doivent d'abord être réduits en
fragments qui, pour ce qui concerne les caractéristiques mécaniques,
physiques et autres, ne peuvent différer que très peu des agrégats naturels
employés commes gros granulats dans le béton.

La présence d'armatures, précontraintes ou non, et de certaines impuretés
comme le plâtre, le bois, les matières plastiques, etc... dans les débris
de béton ou de maçonnerie exige des techniques de séparation appropriées
selon le domaine d'application.

Dans tous les cas, la séparation du béton et des armatures pose un
problème particulier. On cherche à la réaliser au moyen d'installations
de concassage spéciales et de techniques de fragmentation par explosifs.

Tant dans la litérature que dans la recherche en cours, il apparaît que
fabriquer du béton avec des débris provenant de la démolition est déjà
techniquement possible pour certaines applications. On pourrait même, avec
des fragments de béton ancien qui ne seraient pas contaminés par des
impuretés, fabriquer un béton de qualité convenable qu'on destinerait à des
application aux exigences plus sévères.

Reste à savoir si le béton recyclé et le béton fabriqué avec des déchets
do démolition calibrés seront aussi rentables dans un avenir proche et dans
quelle mesure la pénurie de matières premières, les frais de transport des
granulats naturels ou recyclés, l'évolution des coûts de décharge des
déchets et des coûts d'exploitation des granulats naturels, et d'autres

facteurs encore comme la création d'emplois et l'initiative industrielle
privée, vont agir sur les choix possibles et faire opter pour l'emploi
de déchets dans le béton.

La recherch en Europe occidentale porte sur les techniques
de séparation des débris et des fragments de béton armé, sur le développement d'installations pilotes destinées à réaliser cette opération, sur
les aspects technologiques du béton recyclé et sur l'approche économique
du problème.

Etudes nécessaires dans le futur

Dans les années à venir, il sera nécessaire de revoir les connaissances
technologiques et les prescriptions qui concernent le béton en fonction de
la substitution possible des débris ou d'autres déchets aux granulats. Il
faudra aussi étudier l'influence des recyclages successifs sur la qualité
du béton recyclé pour en déterminer las possibilités à long terme.

In ne faut pas cependant se borner à considerer le recyclage du béton
comme un problème isolé mais au contraire s'efforcer d'en maîtriser la
problématique globale pour, em même temps, améliorer ou affiner les
techniques de démolition et tenter d'arriver à une démontabilité totale
ou améliorée des constructions.

ZUSAMMENFASSUNG

Problematik

Es ist damit zu rechnen, dass im kommenden Jahrzehnt einerseits aus
Renovierungsbauprojekten und andererseits aus dem Abbruch von Betonbauten
grosse Mengen von Bau- und Abbruchschutt anfallen werden. Während es sich
bei ersterers hauptsächlich nur um gemischten Schutt handelt, ist es bei
letzterem vorwiegend Beton, der mehr oder weniger mit anderem
Mauerwerkschutt vermischt ist. Eins ist sicher; jedes Objekt und somit
auch ein Bauwerk gelangt früher oder später in den Abfalltrom.
Berücksichtigt man die mittlere Lebensdauer von Betonbauten und die Tat-
sache, dass die Betonherstellung im wesentlichen nach dem Zweiten Weltkrieg
angelaufen ist, stehen wir nunmehr am Fusse des Abbruchschuttbergs. Ohne
Zweifel dürfte bei dem zu erwartenden Schutt Spannbeton eine der
Schuttsorten bilden, deren Verarbeitung schon durch das Vorhandesein der
Bewehrung die weitaus grössten Probleme mit sich bringt.

Vom ökologischen Standpunkt und vom Standpunkt der Versorgung mit
Zuschlangstoffen für Beton her gesehen drängt sich die Wiederverwendung
von Abbruchsschutt auf. Abladeplätze sind immer schwieringer zu finden,
und des Abbau von Zuschlagstoffen wird entweder zum grossen Teil angegriffen
oder aber durch Widerstand gegen das Abtragen blockiert. Wiederverwertung
wäre eine Lösung für beide Probleme.

Es ist daher auch nicht verwunderlich, dass nach Japan und den USA auch in

einigen der am dichtesten bebauten Gemeinschaftsländern die Nutwendigkeit
einer Wiederverwertung von Bau- und Abbruchschutt deutlich geworden ist.
So ist auf Initiative der niederländischen Forschungsorganisation CUR 1977
eine Zussammenarbeit zwischen den Niederlanden, Belgien und Deutschland
zustande gekommen die bisher in den NIederlanden und in Belgien zu
Forschungen über folgende Themen geführt hat: umweltfreundlicher Abbruch
Wiederverwertung und Demontage von Beton usw., zum Beispiel die
Wiederverwendung von Mauerwerkschutt.

Diese Teilstudien sind eng miteinander verflochten. So können
beispielsweise durch die Feststellung neuer Anwendungsbereiche für
widerverwerteten Beton neue Abbruchtechniken erforderlich werden, zum
Beispiel selektiver Abbruch, um Verunreinigungen in rückgewonnenem Beton
zu vermeiden. Andererseits ist demontierbares Bauen eigentlich neue
Strategie, wobei schon im Entwurf der Bauten eine umweldfreudlichere
Abbruchtechnik mit eingeplant wird.

Verarbeitung von Abbruchschutt

Denkt man an die Wiederverwendung von Abbruchschutt als Zuschlagstoff in
neuem sogenannten rückgewonnemem Beton, so muss zunächst der Schutt von
den Abbruchstellen in Bruchstücke zerleinert werden, die sich in ihren
machnischen und physikalischen Eigenschaft und anderen Kennwerten in
vieler Hinsicht nur gerinfügig von den natürlichen Kornungen unterscheiden
dürfen, die als grobe Zuschlagstoffe im Beton gebraucht werden.

Das Vorhandensein von angespannten oder nichtangespannten Bewehrungen und
von einigen Verunreinigungen wie Gips, Holz, Kunststoff usw. im Beton- oder
Bauwerkschutt machen je nach Anwendungsbereich geeignete Trennungstechniken
erforderlich.

In jedem Fall stellt die Trennung von Beton und Bewehrung esi besonderes
Problem dar. Man bemüht sich darum, sie durch die Verwendung spezifischer
Brechanlagen oder durch Stückelungstechniken zu bewerkstelligen, wobei
auch Spengstoff verwendet wird.

Sowohl aus der Lteratur als auch aus den laufenden Forschungen ergibt sich,
dass die Herstellung von Beton mit Abbruchabfällen für bestimmte
Anwendungen schon technisch durchfürbar ist. Es scheint sogar durchaus
möglich zu sein, Beton angemessener Qualitat mit Bruchstücken nicht

verunreinigten alten Betons herzustellen, der für hochwertige Anwendungsbereiche geeignet ist.

Es bleibt noch die Frage, ob rezylierter Beton und mit gestückeltem Abbruchschutt hergesteller Beton auch in naher Zukunft ökonomisch

vertretbat ist und inwieweit die Rohstoffknappheit, die Transportkosten für natürliche und wiederverwerbete Granulate, die Entwicklung der Abladekosten für Schutt, die Arbeitsbeschaffung und industrielle Initiative, die Attraktivität der Verwendung von Schutt in Beton beeinflussen werden.

Die heutige Forschung in Westeuropa konzentriert sich auf Trennungs-techniken für Schutt und bewehrte Abbruchreste, auf die Entwicklung von Pilotanlagen für die Trennungsverfahren, auf die betontechnoligischen Aspekte von rezykliertem Beton oder Abbruchbeton und auf eine möglichst wirtschaftliche Lösung des Problems.

Forschungsbedarf

In den kommende Jahren wird es notwendig sein, die vorhandenen betontechno-logischen Kenntnisse und Vorschriften als Funktion einer etwaigen Substitution der Zuschlagstoffe durch Schutt oder andere Abfallstoffe zu betrachten. Es ist ferner angebracht, den Einfluss der mehrfachen Wiederverwertung auf die Qualität des rezyklierten Betons zu untersuchen um seine langfristige Festigkeit festzustellen.

Gleichzeitig darf das Problems der Wiederverwertung von Schutt nicht als alleinstehende Frage betrachtet werden, sondern muss global von der gesamten Problematik her angepackt werden, und zwar durch gleichzeitige Verbesserung oder Ampassung der Abbruchtechniken une eine vollständige oder verbesserte Demontierbarkeit der Bauten.

Construction like all other industrial activities generates waste products which arise at various stages; during the extraction of the raw materials, during the transformation of the latter into building materials or elements and finally during the construction of buildings and other construction work. Once the technical or useful lifespan of a construction has come to an end it is, just like any other object, destined to find its way into the waste cycle as rubble as the result of demolition, or of pulling down or dismantling for renovation and modernization work.

The rubble from renovation sites is usually of a mixed character, whereas the demolition of concrete structures creates for the most part concrete rubble "contaminated" to a greater or lesser extent with other materials such as plaster, wood, paint, pieces of brick, ... and of course reinforcement elements.

Predictions concerning the quantity of demolition waste we can expect to find ourselves faced with in future years vary widely, a not unimportant reason for this divergence being the differing evaluations of the lifespan of constructions on which the estimates are based.

Although the estimates are not in close agreement, it is generally accepted that a considerable increase in demolition rubble will take place during the coming decade. An important part of this will consist of concrete rubble.

One of the estimates concerning concrete rubble is based on the statistics for cement production from 1920 till present, taking into account the fact that 85 per cent of the cement was used for concrete production and assuming a cement content of 330 kg/m^3 (curve A in fig. 1).

If we then accept as a rough approximation that the average lifetime of a concrete construction is 50 years, with some structures lasting only 30 years while others remain standing after 70 years, we obtain a curve (curve B in fig. 1) which provides us with a tentative estimate for the quantity of concrete rubble which will be generated during the coming years and decades.

One can also add the noteworthy fact that in 1975 the total quantity of concrete in the EEC countries was estimated to be $\pm$ 14 billion tons ($\pm$ 6 billion m^3).

Tipping spots for rubble are increasingly difficult to find and will in future have to be sought further and further from the built-up areas in which the demolition waste arises. Furthermore reinforced concrete rubble is, due precisely to its reinforced nature, one of the most difficult types of rubble to process. In a great deal of countries or regions the sources of supply for many of the raw materials used in the construction industry, and aggregates for concrete in particular, are already exhausted or blocked due to the growing opposition to excavations.

Recycling of building and demolition waste materials could provide a solution to both these problems.

In what follows we shall limit ourselves for the main part to a consideration of concrete rubble, which promises to be of increasing importance in the near future (see fig. 1).

Following the USA and Japan, a number of European countries (GB, NL, F, D, B) have in recent years devoted particular attention to the question of the recycling of concrete rubble. A three country research project was set up by the Netherlands, Belgium and Denmark following the initiative of the Dutch research organization CUR-VB. The project was based on three strongly interrelated themes, comprising : environmentally favourable demolition methods; buildings which can be dismantled, and recycled concrete.

Demolition methods may be influenced by a decision to recycle the rubble. Selective demolition may well be necessary in order to avoid contamination of the recycled concrete by certain impurities, or in order to ensure that the recycled material is suitable for a particular field of application.

The project concerning buildings which can be dismantled dealt with the design of construction joints which would make partial or total dis-mantling possible when a structure had outlived its useful lifetime. This can be considered as a design approach which incorporates an environmentally acceptable demolition method.

After a building has been dismantled the individual elements can either be reassembled for other purposes with, or, without an alteration in function, or they can be processed to create fresh supplies of raw materials, as in the case of recycled concrete, the subject of the third sub-project.

When the reuse of demolition waste materials as aggregate for new "recycled" concrete is considered, then the rubble from the demolition site must first of all be transformed into fragments. The mechanical, physical and other characteristics of these fragments may in many respects vary only slightly from those of the natural materials which are used as coarse aggregate for concrete.

The presence of reinforcement, whether or not prestressed, and of various impurities such as plaster, wood, plastics, etc., in the concrete and brick rubble makes it necessary to use separation techniques adapted to the intended field of application.

Whatever the context, the separation of concrete from its reinforcement poses a particular problem. Attempts are being made to bring about this separation by the use of existing or modified breakers and through fragmentation methods based on the use of explosives.

Fragmentation of demolition concrete

At first explosive charges placed in holes drilled for the purpose were used for the fragmentation of reinforced concrete elements from demolition sites. To avoid the expense involved in drilling holes trials were then made with explosive charges placed on the element concerned. With both techniques it was found that the fragment particle size could be easily controlled by varying the strength of the charge.

The ultimate aim is to use existing or slightly modified rubble breaking installations, breakers for reinforced concrete and fragmentation with explosives in plants designed for the recycling of construction and demolition waste material.

Opinions are still divided as regards the location and scale of such processing plants. Should huge plants be built near large urban centres, in disused or operating stone quarries, near concrete mixing plants, etc., - or by contrast should a number of smaller recycling plants, which could be displaced if necessary, be provided ?

Breaking installations can certainly be moved. However recycling plants where explosive charges were used would have to be fixed establishments, preferably underground or at least in heavily shielded galleries.

Demolition concrete aggregate, produced by means of breakers or explosive charges, consists of the original coarse aggregate used in the preparation

of the "mother concrete", combined with the hardened cement-sand mixture
of the mother concrete.

Thus the porosity of such demolition concrete aggregates corresponds to
that of the mother concrete or the demolition concrete, and is as a
result greater than that of natural aggregate. This factor can influence
the workability of the recycled concrete.

It seems that aggregate derived from demolition concrete is not as strong
as natural aggregate. The lower the strength of the aggregate, the
greater the extent to which the strength of the hardened concrete
depends on this constituent. Consequently the strength of the original
concrete, i.e. the mother concrete has a marked influence on the strength
of the recycled concrete.

Both the specialist literature and the current research findings bear
witness to the fact that the production of concrete using demolition
rubble is already technically feasible for certain fields of application.
It even seems probable that high quality concrete suitable for structural
application can be produced with fragments of uncontaminated old concrete.
The loss of strength is generally limited to between 10 and 20 per cent
compared to an analagous concrete based on natural aggregate.

The modulus of elasticity, an important factor in deformation behaviour,
has been found to be 20 to 30 per cent lower in recycled concrete.

A difficulty which can arise with recycled concrete and which should not
be underestimated is the presence of impurities in the mother concrete.
This contamination can take various forms :

    a.   The concrete can be subjected to chemical influence during
            its lifetime as a result of contact with chemical substances
            e.g. in chemical factories, industrial purification plants,
            or through prolonged contact with ground or sea water.

    b.   The original concrete can also contain impurities in the
            form of pieces of brick, wood, glass, metal, plastic,
            plaster, etc.

Should doubt exist concerning the chemical influences to which the mother
concrete may have been exposed, it is generally speaking uneconomical
under present conditions to carry out a thorough examination of the
rubble. In such cases recycling is no longer feasible.

A large number of the second type of impurities are only harmful in the sense that they are weaker than concrete, and since they are rarely present in more than small quantities they do not exert a great influence.

Plaster is, however, an important exception. It is by nature particularly expansive, and has a detrimental effect if present in excessive quantity in the recycled concrete.

Naturally the quality of the mother concrete or demolition concrete also has a marked influence on the characteristics of the recycled concrete. An idea of the influence of the type of cement originally used, of the cement content, of the type of aggregate, etc., was obtained by recycling 30 types of well-defined 15 year old mother concrete, fragmented for the purpose by the placing of explosive charges.

Some of the results of this study are presented in figs. 3 and 4.

Economic aspects

An important question remains as to the economic feasibility in the near future of recycled concrete and of concrete prepared from graded demolition rubble. And it remains to be seen to what extent shortages of raw materials, transport costs for natural and recycled aggregate, trends in tipping costs for rubble, trends in quarrying costs for aggregate and other constituents, the provision of employment, and industrial initiative will influence the attractiveness of the use of rubble as aggregate for concrete.

At present research in Western Europe is focused on separation methods for rubble and reinforced material generated by demolition activities, on the development of pilot separation plants, on the concrete technology aspects of recycled concrete or rubble concrete and on the economic side of the problem.

In the coming years it will be necessary to review the existing state of the art in concrete technology and to review the regulations and standards applying to concrete in the light of the possibilities offered by the replacement of aggregate by rubble or other waste material.

Finally recycling should not be seen as an isolated question, but instead one should seek an overall strategy for the whole problem area by improving or adapting demolition methods while at the same time constructing buildings and works which can be partially or completely dismantled.

RUBBER RECYCLING BY THE
EUROPEAN RECLAIMERS' ASSOCIATION EURA

Hubertus P.J. Kreemers

European Reclaimers Association

SYNOPSIS

The recycling of cured rubber wastes especially car tyres already started around 1850. At first it was meant to economize their own rubber wastes, but gradually the activity increased to an industrial scale.

During the period 1960-1980 enormous quantities of worn-out tyres have been recycled; USA 4 million tons, EEC 2 million tons. The waste rubber is devulcanized after being treated mechanically. A straining and refining process is necessary to obtain a recycled material according to standard specifications. More and more recycled rubber is used on behalf of its typical processing properties.

The application has been limited and even decreased a.o. due to low prices for natural and synthetic rubbers. The introduction of radial steelcord tyres caused heavy problems in mechanical cracking.

Adding recycled rubber to rubber compounds such as tyre compounds facilitate the processing and curing, and saves energy. In the EEC approximately 140 000 tons of worn-out tyres are recycled yearly and the recycling industry is saving approximately 100 000 tons of crude oil imports.

The European Reclaimers Association aims at a technical development in adjusted tyre-cracking equipment and at the same time wants to stimulate the consumption of recycled rubber. In the EEC about 800 000 tons of worn-out tyres are available per year. At this moment approximately 18 per cent are recycled into "reclaimed rubber" and another 5 per cent into rubbercrumb.

The Association of Rubber Reclaimers propose a number of international activities such as :

1. Organising collection and preparation of waste tyres.

2. Mutual technical research and development of necessary equipment.

3. Improvement of recycling systems.
4. Stimulation of the consumption of both recycled rubber and rubber crumbs.
5. Finding new applications outside the rubber industry.
6. Enlarged activities of the existing recycling industry.

RESUME

Le recyclage des déchets de caoutchouc vulcanisé, notamment le recyclage des pneus de voiture, remonte aux environs de 1850. Au départ, il s'agissait d'économiser les déchets de caoutchouc, mais cette activité s'est progressivement développée au point de devenir une véritable industrie.

Il a fallu attendre la période 1960-1980 pour que d'énormes quantités de pneus usagés soient recyclés : 4 millions de tonnes au Etats-Unis, deux millions dans la CEE. Le caoutchouc de rebut est dévulcanisé après avoir été traité mécaniquement. Il est nécessaire de l'épurer et de l'affiner pour obtenir une matière de récupération conforme aux normes.

On utilise de plus en plus de caoutchouc recyclé en raison des propriétés de traitement caractéristiques de cette matière.

L'application de ce procédé a été limitée et a même diminué notamment en raison du prix peu élevé des caoutchoucs naturels et synthétiques. L'introduction de pneus à carcasse radiale en acier a posé de graves problèmes au niveau du broyage mécanique.
Ajouter du caoutchouc recyclé à des mélanges de caoutchouc tels que ceux utilisés dans la fabrication des pneus facilite le traitement et la vulcanisation et permet de réaliser des économies d'énergie.

Dans la CEE, quelque 140.000 tonnes de pneus usagés sont recyclés chaque année et cette industrie permet d'économiser l'importation de quelque 100.000 tonnes de pétrole brut.

L'association des récupérateurs européens cherche à développer la technique des équipements adaptés au broyage des pneus tout en stimulant la consommation de caoutchouc recyclé.

La CEE dispose de quelque 800.000 tonnes de pneus usagés par an. A l'heure actuelle, 18% environ sont recyclés en "caoutchouc de récupération" et 5% en déchets broyés de caoutchouc.

L'Association des récupérateurs de caoutchouc propose de développer un certain nombre d'activités internationales, par exemple :

1. Organiser la collecte et la préparation des pneus usagés.
2. Collaborer à des travaux de recherche sur la technique et le développement des équipements nécessaires.
3. Améliorer les systèmes de recyclage.
4. Stimuler la consommation de caoutchouc recyclé et de déchets broyés de caoutchouc.
5. Chercher de nouvelles applications en dehors de l'industrie du caoutchouc.
6. Elargir les activités de l'industrie de recyclage actuelle.

ZUSAMMENFASSUNG

Abfälle von vulkanisiertem Kautschuk werden bereits seit 1850 wiederverwertet. Das ziel diese "Recycling" bestand zuerst in der wirtschaftlichen Verwertung eigener Kautschukabfälle, doch weitete es sich nach zu einer Tätigkeit von industriellem Masstab aus.

Erst in der zeit zwischen 1960 und 1980 wurden enorme Mengen abgenutzter Reifen wiederverwertet. In den beliefen sich diese Mengen auf 4 Millionent in der EWG auf 2 Mil t. Der Abfall kautschuk wird nach mechanischer Behandlung devulkanisiert. Um den Standardvorschriften entsêchende Stoffe zu erhalten, ist ein Uberarbeitungs- und Veredelungsprozess erforderlich. Mehr und mehr Recycling-Kautschuk wird aufgrund seiner typischen Verarbeitungseigenschaften verwendet.

Niedrige Preise für Natur- und synthetischen Kautschuk setzen der Wiederverwertung von Altkautschuk Grenzen und haben sogar zu einem Rückgang derselben geführt. Die Einführung von Stahlkord-Gürtelreifen hatte ernstahfte Probleme bei der mechanischen Zerkleinerung zur Folge. Die Hinzugabe von Recycling-Kautschuk zu Kautschukmischungen wie Reinfenmischungen erleichtert die Verarbeitung und das Vulkanisieren und ist energiesparend.

In der EWG werden jährlich rund 140 000 t Altreifen wiederverwertet, und die Recycling-industrie ermöglicht damit eine Einsparung von rund 100 000 t eingeführten Rohöls.

Die "European Reclaimers Association" strebt die technische Entwicklung *von angepassten Einrichtungen zur Zerkelinerung von Reifen an und fördert gleichzeitig den Verbrauch von Recycling-Kautschuk.*

In der EWG fallen jährlich rund 800 000 t Altreifen an. Zur Zeit werden davon etwa 18 % als "Regeneratgummi" und weitere 5 % als "werkleinerter Gummi" wiederverwertet.

Die "Association of Rubber Reclaimers" schlägt eine Reihe internationaler Tätigkeiten vor, u.a. :

1. Organisation der Sammlung und Zubereitung von Altreifen.

2. Gemeinsame technische Forschung und Entwicklung der erforderlichen Einrichtungen.

3. Verbesserung der Recycling systeme.

4. Förderung des Verbrauchs von Recycling-Kaubschuk und zerkleinertem Kautschuk.

5. Auffinden neuer Verwendungen ausserhalb der Kautschukindustrie.

6. Ausweitung der Tätigkeiten der bestehenden Recycling-Industrie.

Recycling vulcanised rubber started around the middle of the last century, with its origins in the reuse of scrap produced during the manufacturing process.

Today recycling has become very extensive. During the last 20 years the quantity of rubber recycled totalled 4 100 000 tonnes in the USA, and 2 100 000 tonnes in EEC countries.

For worn vehicle tyres of which 80 million become available in EEC countries each year, there are five alternative systems that can be adopted. These are :

retreading ;
mechanical size-reduction and cryogenic grinding ;
recycling by devulcanizing into a virgin material ;
conversion into a road surfacing material ; and
pyrolysis.

In principal, all tyres can be processed in one of these ways or another, as most worn tyres suffer only from the cost of some millimeters of rubber from the profile.

As long as a careful selection process is adopted, retreading worn tyres is a practical procedure. However this section of industry suffered from low-pricing of new tyres during last years. The situation could reverse if a 3 mm minimum tread thickness were to be brought into force.

Mechanical size-reduction and cryogenic grinding are used to achieve an extreme reduction in size before reclaiming can start.

Recycling by means of devulcanizing processes is the most complete solution to handle worn-out tyres. In fact all available waste-tyres can so be recycled into valuable and economical products.

Following EEC regulations research for new applications of tyre materials has started. It is now evident that worn tyres could be used for road-surfacing and for noise insulation, for example.

Finally the pyrolysis of waste tyres and even large quantities of non-tyre waste rubbers is a possibility for recycling. Some activities have started recently and the economiçal outcome depends on developments of markets.

Generally speaking all systems of reclaiming rubber aim at a devulcan-ization of the cured rubber wastes in order to plasticize the material and refine it according to standard specifications.

Having started as a low cost replacement of virgin rubbers, the recycled product is now becoming recognised as having specific properties of its own.

During last decades the quality of recycled rubber has improved constantly. Now the industry needs funds to invest in developing economical efficient equipment for handling of steelcord-tyres.

Generally it is ready to solve a part of the environmental problems in the EEC countries by producing important raw materials for use both within and outside the rubber industry.

However between 1960 and 1978 the consumption of recycled rubber has been stable or even reduced, as the table shows.

Consumption of recycled rubber in 1 000 tonnes [*]

	U.S.A.	U.K.	Germany	France	Brazil
1960	276 515	35 200	45 944	29 546	10 116
1970	254 445	23 800	32 319	38 915	20 603
1978	125 828	15 800	12 319	33 918	32 255

Low prices of natural and synthetic rubbers did not encourage producers to use relatively expensive recycled rubbers.

The main qualities produced are :

a. Whole tyre reclaims   from car- and truck-tyres

b. Butyl reclaims   from car innertubes

c. EPT-reclaim   a latest development; in research at several reclaiming factories made from EPT rubber profiles.

These recycled rubbers are generally re-used in the products they are made from.

As a result of the recycling process some technical properties of the original product are partly decreased, but compensated by specific new properties as :

    Shorter mixing times
    Low power consumption
    Low heat development
    Fast processing on extruder and calanders
    Low swelling and shrinkage of the unvulcanized compound
    Faster curing of the compounds.

The greatest part of a tyre consists of products based on crude oil. Thus it can be calculated that the yearly production of approx. 140 000 tonnes recycled rubber in the common market is saving a 100 000 tonne imports of crude oil per year.

To produce one kilo of recycled rubber only an average of one KWH electrical energy is used. On top of this it is nearly impossible to estimate the energy savings in the rubber industry by using recycled rubbers.

---

[*] Rubber Statistical Bulletin

In Holland approximately 4 700 000 worn tyres are recycled each year. The total weight is estimated at 40 000 tonnes. Based on this figure, the total for the EEC could be at 800 000 tonnes. In Holland 62 per cent of the worn out tyres are tipped and we assume, that other countries do not differ very much.

In spite of the fact that rubber wastes, especially worn tyres and tubes, are readily available, and often incur costs to be disposed of, the recycling industry still has to pay a high price for its feedstock. The problem is to find an optimal system of collection and supply. The industry feels that this is a task for the European institutions.

Comparative calculation for a recycled whole tyre versus virgin materials :

1 000 kilos of recycled tyres contain :

		Average per cent	Weight	Virgin value
1.	Oils	20	200 kilos	£ 42,-
2.	ZNO	4	40 "	£ 17,-
	Fillers	2	20 "	£ 4,-
3.	Carbon black	32	320 "	£ 82,-
4.	Rubber	42	420 "	£ 252,-
5.	Energy	1 000 KWH		£ 27,-
			1 000 kilos	£ 424,-

Market price of the recycled rubber	=	£ 260,-
Consumers price advantage	=	£ 164,-

EEC-consumption 140 000 tonnes
Total advantage 140 000 x £ 164,-   = £ 22 960 000,-
                                                  ============

The situation at present is that of the 1 400 000 tonnes rubber produced only ± 17 per cent is being retreaded, recycled or reused via pyrolysis. This means therefore that ± 1 160 000 tonnes of valuable rubber polymer is being dumped, buried or burnt.

The industry feels that an approximate doubling of the present recycled rubber consumption can be achieved without great research efforts on end product and techniques.

The Association of the European Rubber Reclaimers therefore proposed the following actions :

A common market organisation dealing with the collection and the preparation of the required rubber wastes ;

Technical research and development of suitable equipment for cracking steelcord tyres ;

Research for finding adjusted recycling systems ;

Necessary means to raise the performance of our products, to improve the image, to increase the applications and to develop new types of recycled rubber.

Recycled rubbers have to find their way not only in the traditional outlets of rubber products but also in new applications such as road-surfacing, industrial flooring, sports surfacing, acoustic insulation, adhesives and anti-corrosive protection.

THE FUTURE -   WHERE DO WE GO FROM HERE ?

Peter Menke-Glückert

Ministry of the Interior

Federal Republic of Germany

SYNOPSIS

Waste quantities will increase in the forthcoming decade, though the increase will be less than during the past years. The waste problem will continue to gain importance.

Waste management, in particular waste reduction and waste recycling, is being increasingly determined by requirements of raw material conservation and energy saving. Waste management exceeds more and more the field of environmental protection and becomes an integrated issue of an industrial policy taking into account a rational use of raw materials. Production planning and consumer behaviour have to be systematically adjusted to requirements of waste reduction and material recycling.

The direct waste prevention (prevention at source) has to be more emphasized in the future than it is to date. This holds true for the sectors of consumer wastes and production wastes. Only a conscientiously pursued policy of waste prevention will effectively contribute to a long-term saving of raw materials. Developments in the areas of waste prevention and waste recycling should be guided, as much as possible, by principles of market economy. The instrument of voluntary agreements between governments and industry on the achievement of waste management objectives is of special importance in this framework.

Meanwhile, the enforcement of the five directives dealing with the disposal of waste is of particular priority in the European Community in the forthcoming years. The harmonizing process at which the directives aim will only be achieved, if member countries do not only enact the necessary regulatory instruments for waste disposal but also install appropriate facilities and equipment for an environmentally sound waste disposal system.

RESUME

L'augmentation des quantités des déchets, bien que moins forte que dans les années passées, se poursuivra également dans la décennie prochaine. Le problème des déchets devient donc de plus en plus important.

La gestion des déchets, notamment la réduction des déchets à la source et la récupération, est déterminée de façon croissante par les aspects de l'économie des ressources en matières premières et énergie. La gestion des déchets dépasse donc de plus en plus le cadre de la protection de l'environnement et devient une tâche intégrante d'une politique industrielle consciente du problème des matières premières. Tant la planification de la production que les habitudes de consommation doivent être adaptées systématiquement aux nouvelles exigences de la réduction des déchets à la source et du recyclage des matières premières.

Dorénavant, il faut attribuer une plus grande importance à la réduction des déchets à la source que jusqu'à présent. Ceci s'applique simultanément aux domaines des déchets de consommation et de production. A la longue, les matières premières ne peuvent être économisées de façon efficace qu'au moyen d'une politique conséquente en matière de réduction des déchets.

Les développements dans le domaine de la réduction et de la récupération des déchets en tant que tâches intégrantes de l'évolution économique devraient s'orienter dans la mesure du possible vers les critères de l'économie de marché. A cet égard, l'instrument d'accords volontaires entre l'Etat et l'Industrie sur la réalisation d'objectifs en matière de gestion des déchets revêt une importance particulière.

Au cours des années prochaines, la priorité doit être donnée à la réalisation des cinq directives cadres convenues au plan communautaire en vue de l'élimination des déchets. L'harmonisation envisagée dans ces directives en ce qui concerne les exigences de l'élimination des déchets n'est obtenue que lorsque les Etats membres ne créent pas seulement les dispositions légales, mais aussi les installations nécessaires pour l'élimination des déchets.

ZUSAMMENFASSUNG

Die Adfallmengen werden auch im kommenden Jahrzehnt, wenn auch weniger stark als in den zurückliegenden Jahren, ansteigen. Das Abfallproblem gewinnt damit weiter an Bedeutung. Die Abfallwirtschaft, insbesondere die Abfallvermeidung und Abfallverwertung, werden zunehmend von Gesichtspunkten der Rohstoff- und Energieeensparung bestimmt. Abfall- wirtschaft geht damit mehr und mehr über den Bereich des Umweltschutzes hin aus und wird zu einer integrierten Aufgabe einer rohstoffbewussten Industriepolitik. Sowohl Produktionsplannung wie auch Konsumerverhalten sind systematisch an neue Erfordernisse der Abfallvermeidung und Kreislaufführung von Rohstoffen anzupassen.

Die direkte Abfallvermeidung (Vermeidung an der Quelle) Muss künftig grösseres Gewicht als bisher erhalten. Dies gilt gleichermassen für die Bereiche der Konsum- und Produktionsabfälle. Nur über eine konsequent betriebene Abfallvermeidungspolitik kann auf Dauer ein wirksamer Beitrang zue Rohstoffeinsparung geleistet werden.

Entwicklungen im Bereich der Abfallvermeidung und Abfallverwertung sollten sich als integrierte Aufgaben des Wirtschaftsprozesses soweit wie möglich nach marktwirtschaftlichter Kriterien richten. Dem Instrument freiwilliger Vereinbarung en zwischen Staat und Wirtschaft über die Verfolfgung abfallwirtschaftlicher Ziele kommt in diesem Zusammenhang besondere Bedeutung zu.

Mit vorrang gilt es in den nächsten Jahren die auf Gemeinschaftsebene beschlossenen fünf Rahmenrichtlinien zue Abfallbeseitung in die Praxis umzusetzen. Die von diesen Richtlinien verfolgte Harmonsierung der Anforderungen an die Abfallbeseitugung wird nur dann erreicht, wenn die

Der Abfallwirtschaft ist künftig auf Gemeinschaftsebene grösseres Gewicht Beizumessen. Die Lenkungs- und Koordinierungsfunktionen des EG- Ausschusses für Abfallwirtschaft sind zu intensivieren. Dies ist eine wesentliche Voraussetzung für die notwendige Abstimmung der EG- Abfallwirtschaftspolitik zwischen den EG-Mitgliedstaaten, der EG- Kommission und mit anderen Beieichen der EG-Wirtschaft- und Umweltpolitik.

There are many possible ways in which waste disposal policy could develop
in the future. To look at some of them in 1980 involves giving reliable
forecasts on total economic development, further changes in the raw
material and energy market and, not least, the future state of the environment. In view of the short-lived nature of modern political events, no-one
will claim to predict tomorrow's development reliably from today's knowledge.

Yet there are clear indications of the outline conditions which will
determine waste disposal in the next 10 to 15 years. The raw material and
energy aspects will become increasingly important factors of waste
disposal. The resulting economic stimulus will make the avoidance or use
of waste an even more obvious requirement of industrial planning than it is
today.

Environmental requirements support this development. In the past ten years
it was originally the aim to issue environmental and waste disposal legislation and corresponding guidelines at an EEC level. In the next decade
the aim will be to make the regulations more concrete and apply them more
consistently in practice. It is only when the regulations have been
actively put into effect - and we are very far from this situation in some
cases - that many waste producers will realize that the non-production
or utilization of wastes are often better alternatives to waste disposal,
which is subject to strict requirements.

Today 80 to 90 million tonnes of domestic refuse are produced per year in
the EEC. Does this figure represent the end of a development, a turning
point ? Unfortunately not.

Assuming that the long-term trends of previous years will continue over
the next 10 to 15 years, the following scenario is probable. The gross
national product will probably increase in the next decade, in spite of
EEC energy and raw material problems. There will therefore probably also
be an increase in private consumption, resulting in a further increase in
domestic waste.

During this development, consumer demand will probably concentrate on high-
quality products, whereas the demand for short-life goods will decrease
less. On this assumption, we may reckon with an average increase of about
1.0 to 1.5 Wt. per cent per year in domestic waste over the next ten years.
On the other hand an increasing proportion of waste will be consumer goods

made of higher quality materials.

In 1990, therefore, the production of domestic waste in the EEC will probably be 100 million tonnes, i.e. a problem of extraordinary magnitude.

The proportion of paper in domestic waste will increase, the proportion of plastics will decrease and the metal content will remain substantially unchanged.

With regard to particular kinds of waste, the changes will not be uniform. There will probably be an increase in waste salts or sulphuric acid as a result of a further growth in chemical production, but in other sectors, e.g. the metal working industry, there will be no change or possibly a reduction in waste production. Further development in the EEC will also depend on whether legal regulations regarding the disposal of special waste are used effectively to preserve the environment.

One special problem is the disproportionate increase in sludge and dust resulting from water and/or air purification. More particularly, the construction of new coal-fired power stations will result in considerable waste problems in most EEC countries.

This brief scenario shows that the waste situation in the EEC is likely to become worse rather than easier in the next few years. Efficient waste disposal will become more necessary than ever.

It may be asked whether this problem can be solved by halting economic development (a crisis situation), which may result anyway through the serious problems of raw materials and energy supply. The answer is undoubtedly no. Such a development would certainly result in a reduction in waste, but probably also in a relaxation of efforts to preserve the environment coupled with the endangering of existing processing activities.

Waste disposal, with its problems and tasks, is a mirror image of economic and social life. In view of this background, it would be presumptuous to expect waste disposal policy by itself to provide a solution to the urgent questions of saving raw materials and protecting the environment. Effective progress in these sectors will be possible only if the general principles of thrift and sparing use of resources are more firmly accepted in state, industrial and private planning.

At present we are far from this objective, in spite of many protestations to the contrary. It will be of no use to later generations if we prophesy a shortage of raw materials but are incapable of effectively counteracting this threat today. In future years, the task will be to introduce a fundamental change of attitude in this connection.

At the present progress in waste disposal policy often encounters difficulties because the individual interests of certain branches of industry and short-term expectations of profit are valued above the goal of long-term protection of reserve raw materials and the environment. Drastic rethinking is essential here, helped mainly by reformulated state outline planning. The required long-term planning will not mean a planned economy but the guidance of commercial planning towards new goals, with the following important points :

> opening up the market by improved information about waste substances and their possible uses ;

> long-term supply and acceptance contracts for waste materials for re-use ; "conventional" waste disposal system, as in the paper sector, must not only be preserved but adapted to new requirements regarding the preservation of raw materials and the environment ;

> development of the use of waste as part of public waste disposal by better cooperation between state and private industry ;   and

> closer cooperation within industries with a view to recycling between sectors.

The development of new technical methods of avoiding and using waste is undoubtedly a problem central to all waste disposal efforts. These developments require time, more particularly with regard to their subsequent practical application. Before 1985, for example, it will be impossible to make a universally binding prediction about the prospects of large-scale use of domestic refuse in refuse-sorting plants. There is thus a need for prompt, preventive action today to ensure success tomorrow. The state and industry are therefore called upon to intensify promotion of relevant research and development projects. In view of the variety of problems, the aim should be a rational distribution of labour between EEC Member States.

Promising starts in the avoidance or use of waste should not hide the fact that the elimination of waste in the next few years will be subject to serioues problems and difficulties, at least locally. This is suggested not only by the quantity of waste in absolute terms, but also by the present increase in concentration of harmful substances in waste as a result of other environmental measures. This development is all the more critical in that the existing supply of waste disposal installations is quite inadequate, particularly for dangerous waste.

Any improvement in this situation will require special efforts from EEC Member States, in that the siting of new waste disposal installations is encountering great difficulties for technical and geographical reasons, and particularly because of public opposition.

The consumer, through his demand and everyday behaviour, will always play a central part in the development of waste disposal policy. At present and in practice, this influence is small and has been progressively restricted in recent years. The consumers' freedom to reduce waste is progressively restricted under the pretext of "rationalization".

This trend must be appreciably reversed in order to prevent a further increase in waste from consumers. For example, it is quite unreasonable that an economically and ecologically sensible system such as returnable packaging should be progressively broken up and the consumer should be increasingly forced to use throw-away containers. In this respect, it is not enough to indicate the alternative of "recycling", since this method is usually neither advantageous to the consumer nor economical of resources. It cannot be in the interests of waste disposal policy to take one step forward and two steps back.

Demands are also made on the consumer with regard to his quality claims on certain products. Present-day advertising is still oriented contrary to waste disposal requirements. As before, there is insufficient information on the consumption of energy and raw materials by certain products and their effects on the environment.

The consumer is given too little information about the detailed cost of many decisions to buy. In the next few years it will be our task to change this situation and at last enable the citizen to make an effective contribution to waste reduction. He is ready to do so but is ill-informed about specific possibilities.

This conclusion, with few exceptions, also applies to the participation of citizens in the use of waste. Our industrial society cannot permanently ignore the increasing waste consciousness among consumers : their potential good-will must be used. Industry must recognize the citizen not only as a prospective consumer but also as a permanent partner in ensuring a supply of raw materials.

The use of waste does not relieve the raw material and environmental situation unless accompanied by intensive efforts to avoid direct waste. A further exponential increase in raw material consumption cannot be compensated by recycling, however successful. This sober mathematical law clearly shows that recycling cannot be a substitute but must be a complement to efforts to avoid waste.

The producer's task is to make greater allowance for the situation in planning his products. The use of low waste technologies, the prolongation of the life of products and the requirement for multiple use of products are suitable and necessary beginnings, which effectively take account of economy in resources without interfering with the qualitative growth of the economy.

In order to save raw materials and energy, it will also be necessary to rethink the role of labour and services in the economic process. In the past, cheap raw materials and energy have suppressed a number of operations and services which economized in raw materials.

The production and use of new products with simultaneous consumption of primary raw materials and production of new waste was, and in many cases is, given preference over more labour-intensive processing, repair and further use of existing products or secondary raw materials.

Without wishing to arouse excessive expectations in this connection, we think it advisable, even allowing for labour-market policy for producers and workers in future to recall the almost forgotten principles of economy in materials which used to be applied. There is no doubt that greater emphasis on services in industrial development will offer possibilities, at least in some sectors (e.g. in greater use of repairs) of reducing the problem of waste.

Waste is not usually locally concentrated but widely scattered over an area. The resulting problems regarding transport distances and full use of plant capacity require special development of decentralized methods of waste use. The future success of waste disposal policy will not necessarily depend on the amount of investments to be made but also on the flexible adaptation of existing means to local and regional requirements. This means that small and medium-sized undertakings will have a central function in the further development of waste disposal policy.

The solution of raw material and environmental problems will undoubtedly pose new survival tests in coming years on the commercial system practised in EEC countries. In the interests of the recognised advantages of our free economic system, these tests must be passed by the use of substantially market-oriented methods. Admittedly this basic requirement, like the social question, implies that industry and consumers alike will be ready to accept new outline conditions and aims which take into account the supply situation and are more long-term than in the past.

The central task of national waste policy in coming years will be to set clear objectives to industry regarding the avoidance or use of waste, as has already occurred in the energy sector. The task of industry, for its part, will be to make the most efficient use of its freedom of action in its own interest, allowing for its responsibility to the general public.

The precedence of private initiative and responsibility must be maintained, even in the solution of waste disposal problems. The extension of state bureaucracy and planning is not the most promising way of solving practical problems regarding the avoidance and the use of waste. This should be remembered particularly by the governments of EEC Member States and by Community institutions.

Against this background, voluntary agreements between state and industry are of central importance in the further development of waste disposal policy. In this connection, the main function of national policy will be to define clear objectives regarding the avoidance or use of waste. The function of industry will be to make effective use of its existing freedom of action, under its own responsibility and without bureaucratic control. The main advantage of this distribution of responsibility is flexible adaptation of individual measures to new technical and economic developments.

Voluntary agreements between state and industry, and their practical application, pose new and in some cases heavy requirements on both partners.

The main requirement of national policy is to depart from its process of arriving at mostly short-term decisions. Waste disposal policy should not only concentrate on balancing interests between commercial partners and on short-term commercial arrangement. National policy must also have the courage to act against individual market interests and set long-term goals, if required by the environmental and raw material situation. Cooperation with industry does not exclude confrontation with some participants.

Industry must also learn the use of voluntary agreement. Such agreements, as a fundamental part of waste disposal policy, will be successful only if they are observed by all industrial parties. A situation in which some keep to agreements whilst others ignore official policy and thus obtain commercial advantages, is politically unacceptable. It would endanger the basic concept of a substantially market-oriented waste disposal policy.

Some EEC Member States such as France and the Federal Republic of Germany are already successfully applying voluntary agreements to waste disposal policy. Their experience in solving the packaging problem should stimulate an extension of corresponding agreements to other sectors, e.g. elimination of heavy metals from domestic refuse, increased use of waste paper and the reduction of toxic sludge from the treatment of metal surfaces.

The main aim should be to obtain comparable agreements in the various EEC Member States. In this connection the aim should be that agreements in one country are respected by neighbouring countries and are not circumvented by trans-frontier import and export trade.

With regard to subsequent development, it would be wrong to regard waste disposal as a purely public problem and the use of waste as a purely private task. This clear delimitation of responsibility is not an acceptable solution. In many cases the use of waste is a part of waste disposal policy and vice versa. In view of this close connection, the public sector must show more understanding and consideration for economic factors. On the other hand, private industry, in its processing activities, must attach more importance than previously to a reliable waste disposal system.

The European Economic Community, in two environmental action programmes and five directives, has already indicated aims and measures for a future oriented waste disposal policy. The EEC Conference from 17 to 19 June 1980 is a visible expression of initiatives and the importance of this branch of Community policy.

However, the abundance of new activities should not disguise the fact that
EEC waste disposal policy is still in a preliminary phase. Most EEC
directives, e.g. on waste oil, waste and elimination of dangerous waste,
have been incorporated only partly in the national legislation of Member
States. The extent to which this environmental legislation has been put
into effect is still far from satisfactory, particularly in the case of
toxic waste. There is thus an urgent need to develop and apply uniform
elimination criteria, and devise suitable disposal equipment in all EEC
Member States.

As part of waste disposal policy in the next few years, it will be
necessary to :

> work out a general EEC plan for waste elimination on the high
> seas; and check regulations on the import and export of toxic
> waste in EEC Member States. The control of the flow of waste
> should not be limited by national boundaries.

With regard to the avoidance and use of waste, emphasis will in future be
placed on the development of a Community policy on waste paper and
packaging. Additional work must be intensified and rationalised. The aim
of this work is not necessarily to draft and issue new directives but,
instead, an attempt should be made by close cooperation with the relevant
industry to obtain flexible solutions allowing for the special situation
of particular Member States. Allowance must also be made for the
possibility of agreements between sectors covering the entire EEC.

EEC waste disposal policy implies close collaboration between EEC Member
States. This will require a further extension of concerted action in the
field of research and development, improved exchange of information and
experience between EEC Member States, and increased guidance and
coordination by the EEC Committee on Waste Management.

According to present predictions, the known stocks of some important
mineral raw materials will last for only 20 to 40 years. The oil problem
is becoming more acute. This sober assessment will force the Community to
act, since it is highly dependent on raw material imports. There is no
alternative to a policy of thrift. The European Parliament, the
governments of Member States and the EEC Commission are called upon to act
in accordance with these findings.

EUROPEAN CONFERENCE ON WASTE MANAGEMENT

FINAL STATEMENT

The Conference reaffirmed the continued and growing need for effective waste management aiming in particular at appropriate product design and effective waste prevention and resouce saving politics.

To this end it emphasized the nedd for national and regional governments to give specific and adequate attention to problems related to waste generation and disposal, especially for hazardous waste and for the Community to initiate action, coordinate and harmonise national measures and objectives through all its branches.

The immediate needs for the next few years appear to lie in the areas of:

- hazardous waste
- packaging
- greater use of recycled materials
- energy from waste

The Conference called for the Community to further integrate, in the next decade, waste management strategies in the overall economic development and regional planning, for instance, by initiating and promoting demon- stration projects and training and education programmes aiming at an increased industry and consumer awareness.

At the same time, further means of specialised communication and exchange of information and experience should be developed at the appropriate level.

Based on the usefulness of the exchange of information and experience this conference has given it was agreed that a future conference should be held in two or three years time, reviewing the progress achieved in the meantime.

THE FUTURE - WHERE DO WE GO FROM HERE ?

Report of the concluding debate - Chairman Dr Robert L.P. Berry

Following three days of continuous bludgeoning by approximately 100,000 words on subjects as diversified as recycled rubble and the chemical content of livestock excrement delegates could be forgiven for feeling somewhat stunned on reaching the concluding debate.

However the purpose of their final task was clear. It was to strive to draw up clearly expressed conclusions on issues that could seem at that stage to be confused rather than just complex.

And it can be said that interesting points did arise during the final discussion, covering both specific subjects as well as the dawning of a clear light in which to view the discipline of waste management as a whole.

In other words the objectives that the European Commission had in mind on conceiving the whole ambitious project were finally realised - in substantial part at least.

One of the several specialist subjects to be covered perhaps epitomised the positive attitudes arising throughout both the discussion and the conference. That was Dr Lisa Pavan's (Commission) proposal for an industry prize for an ecologically designed beverage container.

"We have seen many beautiful designs from the aesthetic point of view, but not one taking into account ecological parameters", she said.

Her proposal, that fell within the sensitive and fraught field of beverage containers, had followed a similar idea put forward by arch ecologist and powerful debater, Mr Tom Burke, of the Friends of the Earth Society. Mr Burke had called for a European award for a design that would minimise the waste of material.

Unfortunately more than prizes for good behaviour is needed to achieve major policy changes.

They are no more useful than general exhortations to industry to behave in a virtuous manner.

One typical reaction to these exhortations came from feet-on-the-ground industrialist Mr E. Vaughan Southam of Dickinson Robinson, UK. He expressed the overall reaction with the words : "I cannot see how industry can tell ITS customers what to buy."

The implication was clear. No industrial concern can survive if it moulds its policies to respond to exhortations.

And these feelings bring into relief the enigma that surfaced frequently throughout the conference. That is, whatever the desires of individual members, industry's fundamental motivation has to be to operate in competition with itself, at a profit. This does not usually coincide with the quite different objectives of resource recycling.

Very often, according to Miss M.A. Lund, of the Atomic Energy Establishment at Harwell, recycling technologies were already known in various branches of industry. Pilot plants had proved successful.

But despite these facts, techniques often failed to be adopted. This was because profitability was either marginal, or non-existant.

Goodwill could be expected to motivate only in major environmental issues, she stated. The example she then quoted was the favourable response by industry to the fluorocarbon/ozone layer issue.

What means exist, asked Miss Lund, to encourage industry to adopt recycling techniques in marginal cases?

Certainly any call for subsidies was not vocal among the British. These could involve massive expenditure with very little result, was how Mr G.M. Wedd, of the UK Department of the Environment, expressed this feeling during the concluding debate.

The idea that the attitude of industry should be to think of money spent on resource recycling as a form of investment in new production processes was offered up by Tom Burke.

A warm response was drawn by Prof Paolo Schmidt di Friedberg, of Montedison, Italy, who compared the value of recycling with certain "free facilities" such as the pleasant rural landscape, which everyone judges to have a value which goes beyond the strictly financial one.

"This is an interdisciplinary field and there is a need for a major evaluation of it ... what we (industry) are asking for are clear cut goals, so that we know what direction you want us to go in", he said.

It was not a question of whether waste management policy was necessary, but how to go about achieving the best means, was something of the same thought, expressed differently, by Mr Peter Menke-Glückert.

As the debate got round to looking for answers to the questions so often posed several ideas began to flow. Good communication took a high priority. There was a need for frequent discussion between governments, industry and consumers, said Mrs Aloisi de Larderel, of the French Ministry of the Environment.

It was anomalous, therefore, that employment, which needed encouragement by governments, happened to be a principal source of taxation. At the same time raw commodities, involving minimum eployment, and much of that outside the community countries, escaped any particular taxation.

Officialdom should not on the one hand exhort society to conserve imported raw materials when at the same time it provided a hefty financial penalty to industry involved in doing exactly that. The solution was, of course, a fundamental rethink of certain taxation policies, thought Mr. Oxborough.

Another means of encouraging the recycling of waste is research aimed at improving the relevant technologies. Though not a mjor discussion subject during the final debate, research of this kind had been given ample coverage throughout the conference.

The same can be said for publicity, which has already played a successful role in encouraging the public to return bottle for recycling. Perhaps a similar public awareness campaign could one day be equally effective in achieving acceptance of grey recycled paper for household and hygiene use.

Assuming that all pertinent issues can rightly be attributed to conclusions arrived at by the European Conference on Waste Management held at Wembley in June, 1980, then the conference can be said to have stated that if attention were given to the several subjects of :

- purposeful economic incentive structures
- research, both government and industry backed
- harmonisation of legislation throughout the EEC trading
  area to avoid.
- continued communication and discussion at all levels

and   - public awareness campaigns on certain issues

then, efficient recycling of raw material resources will be greatly encouraged.

As a consequence EEC balance of payment figures and social benefits throughout EEC member countries will be improved in both the medium and long term future. In addition risks of economic catastrophe following leaps in the prices of commodities, as happened with oil, will be lessened.

Jeremy Woolfe

TITLES OF SPEAKERS AND CHAIRMEN

Dr Robert L.P. Berry
Director
National Anti-Waste Programme
Great Britain

Michel Carpentier
Head of the Environment &
Consumer Protection Service
Commission of the European
Communities

Yvan Cheret
National Federation of Waste
Activities
France

Dr Vincent A. Dodd
Dublin University
College
Ireland

Dr J.H. Erasmus
Ministry of Public Health &
Hygiene
The Netherlands

Dr James T. Farqhar
Albright & Wilson
Great Britain

Ing. Edmund Fassotte
Directorate General for
Research, Science and
Education, Commission of
the European Communities

Dr Gian L. Ferrero
Directorate General for
Research, Science and
Education
Commission of the European
Communities

Michel de Grave
Social & Economic Committee
of the European Communities

Dr P. l'Hermite
Directorate General for Research
Science and Education
Commission of the European
Communities

Dr Werner Hoffmann
European Container Glass Federation
& Gerresheimer Glas AG
Federal Republic of Germany

Dr George Holzhey
Cepac Waste Paper Committee
& Haindle Paper GmbH
Federal Republic of Germany

Dr Leslie O. Hopkins
Institute for Industrial Research
and Standards
Ireland

Rt. Hon. Tom King M.P.
Minister for Local Government &
Environmental Services
Great Britain

Léon Klein
Environmental & Consumer Protection
Service
Commisssion of the European
Communities

Hubertus P.J. Kreemers
European Reclaimers Association &
Vredestein Industrial Products BV
The Netherlands

Mrs Jacqueline Aloisi de Larderel
Ministère de la Culture & de
l'Environnement
France

Peter Menke-Glückert
Ministry of the Interior
Federal Republic of Germany

Dr Carlo Noto La Diega
Sorain – Cecchini S.p.A.
Rome

Dr H. Ott
Directorate General for
Research, Science & Education
Commission of the European
Communities

Carlo de Pauw
Science & Technical Centre
for Building
Belgium

Dr Lisa Pavan
Environment and Consumer
Protection Service
Commission of the European
Communities

Mogens Rasmussen
A/S Vølund
Denmark

Dr Benno W.K. Risch
Environment and Consumer
Protection Service
Commission of the European
Communities

Prof Paolo Schmidt de Friedberg
Montedison
Italy

Mr John K. Smout
Great Britain

Ben van der Weerden
S. Levison B.V.
The Netherlands

Leopold Van Wambeke
Directorate General
for Research, Science and
Education
Commission of the European
Communities

Jean-Paul Vellaud
National Waste Disposal and
Recovery Agency
France

Dr Gerrit H. Vonkerman
European Environment Bureau

Dr Ing. Paul Weber
Ministry of the Environment
Luxembourg

Bernd Wolbeck
Federal Ministry of the Interior
Federal Republic of Germany

ADDENDUM

EEC LEGISLATION ON WASTE MANAGEMENT

Directive on waste - Council Directive no 75/442/EEC of
15 July 1975 - O.J. L 194 of 25 July 1975

Toxic and dangerous waste - Council Directive no 78/319/
EEC of 20 March 1978 - O.J. L 84 of 31 March 1978

Disposal of waste oils - Council Directive no
75/439/EEC of 16 June 1975 - O.J. L 194 of 25 July
1975

Disposal of PCB's and PCT's - Council Directive no
76/403/EEC of 6 April 1976 - O.J. L 108 of 26 April
1976

Waste from titanium dioxide industry - Council
Directive no 78/176/EEC of 20 February 1978 - O.J. L54
of 25 February 1978.